Glencoe McGraw-Hill

Math Triumphs

Foundations for Algebra 1

McGraw Hill Glencoe

Photo Credits

iv Alamy; **v vi** Getty Images; **vii** Ilene MacDonald/Alamy; **viii** CORBIS; **1** PunchStock; **2–3** Daniele Pellegrini/Getty Images; **7** FlatEarth Images; **13** Dale Wilson/Getty Images; **26** RubberBall Productions; **29** Alamy; **33** Getty Images; **40** Jacques Cornell/The McGraw-Hill Companies; **45** Getty Images; **46–47** Robert Michael/CORBIS; **59** Getty Images; **68** SuperStock; **69** PunchStock; **73** Terry Vine/Getty Images; **74** Getty Images; **78** Brand X Pictures; **79 81** Getty Images; **82–83** Ariel Skelley/Getty Images; **93** Getty Images; **97** Nancy R. Cohen/Getty Images; **98** Getty Images; **104** Alamy; **108** Getty Images; **112** Janis Christie/Getty Images; **118** Noel Hendrickson/Getty Images; **123** CORBIS; **124** Getty Images; **129** CORBIS; **134–135** Alamy; **139** Dennis MacDonald/Alamy; **140** Ei Katsumata/Alamy; **144** PunchStock; **145** Ariel Skelley/Getty Images; **154** CORBIS; **168–169** Ariel Skelley/Getty Images; **173** CORBIS; **174** PunchStock; **179** Ryan McVay/Getty Images; **180** David L. Moore/Alamy; **183** Courtesy of Kings Island; **184** Maxime Laurent/Getty Images; **193 199** Getty Images; **207** CORBIS; **216–217** Tyler Barrick/Getty Images; **251** George Doyle/Getty Images; **257** The McGraw-Hill Companies.

*The **McGraw·Hill** Companies*

Glencoe

Send all inquiries to:
Glencoe/McGraw-Hill
8787 Orion Place
Columbus, OH 43240-4027

ISBN: 978-0-07-890846-0
MHID: 0-07-890846-9

Math Triumphs: Foundations for Algebra 1
Student Edition

Printed in the United States of America.

2 3 4 5 6 7 8 9 10 066 17 16 15 14 13 12 11 10 09

Chapter 1	Numbers and Operations
Chapter 2	Decimals
Chapter 3	Fractions and Mixed Numbers
Chapter 4	Real Numbers
Chapter 5	Measurement and Geometry
Chapter 6	Probability and Statistics

Contents

Chapter 1 Number and Operations

Denver, Colorado

Chapter 2 Decimals

Albuquerque, New Mexico

Contents

Fractions and Mixed Numbers

Niagara Falls, New York

Real Numbers

Charleston, South Carolina

Contents

Chapter 5 Measurement and Geometry

St. Augustine, Florida

Contents

Chapter 6 Probability and Statistics

Mount Rainier National Park, Washington

Chapter 1

Numbers and Operations

How high is the mountain?

The elevation of an object is measured by the distance it is above or below sea level. For example, the elevation of a mountain is the distance its peak is above sea level. To represent distances above or below sea level, you can use integers.

STEP 1 Chapter Pretest Are you ready for Chapter 1? Take the Chapter 1 Pretest to find out.

STEP 2 Preview Get ready for Chapter 1. Review these skills and compare them with what you will learn in this chapter.

What You Know	What You Will Learn
You know how to add. **Example:** $5 + 4 = 9$ **TRY IT!** 1 $2 + 5 =$ ____ 2 $3 + 5 =$ ____ 3 $4 + 7 =$ ____ 4 $6 + 6 =$ ____	*Lesson 1-3* You can use a number line to add integers. $-3 + -2 = -5$ 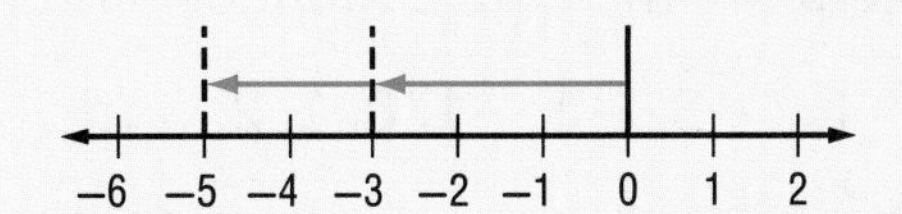
You know how to subtract. **Example:** $12 - 4 = 8$ **TRY IT!** 5 $14 - 8 =$ ____ 6 $10 - 4 =$ ____ 7 $9 - 6 =$ ____ 8 $13 - 5 =$ ____	*Lesson 1-4* You can use algebra tiles to subtract integers. $-5 - 4 = -9$ 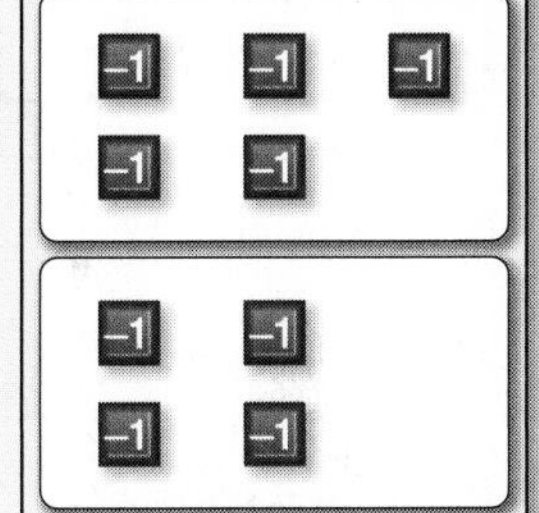
You know how to multiply. **Example:** $5 \cdot 9 = 45$ **TRY IT!** 9 $6 \cdot 6 =$ ____ 10 $8 \cdot 4 =$ ____ 11 $10 \cdot 4 =$ ____ 12 $7 \cdot 5 =$ ____	*Lesson 1-5* To multiply integers, multiply their **absolute values**, then write the sign of the product. **Find $-3 \cdot 4$.** $\lvert -3 \rvert \cdot \lvert 4 \rvert = 3 \cdot 4 = 12$ So, $-3 \cdot 4 = -12$.

Lesson 1-1

Number Groups

KEY Concept

Every number is part of one or more sets of numbers.

Counting numbers are numbers used to count. They are sometimes called natural numbers.

{1, 2, 3 ...}

Whole numbers are the counting numbers and zero.

{0, 1, 2, 3 ...}

Integers are the whole numbers and their **opposites**.

{..., −3, −2, −1, 0, 1, 2, 3 ...}

On a number line, the numbers to the right of zero are positive numbers. The numbers to the left of zero are negative. Zero is neither positive nor negative.

Opposites are numbers that are the same distance from zero on a number line, but in opposite directions.

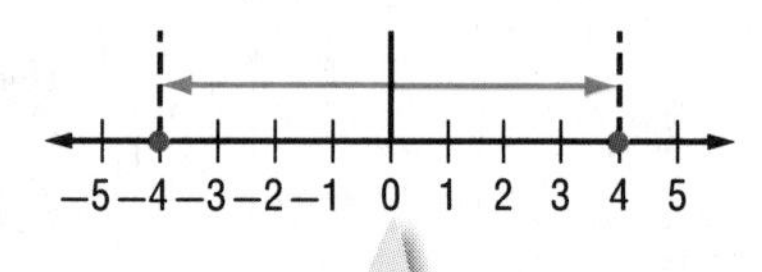

4 and −4 are both four units from zero.

VOCABULARY

counting numbers
numbers used to count objects

integers
whole numbers and their opposites

natural numbers
another name for counting numbers

opposites
two numbers that are the same distance from zero on a number line in different directions.

whole numbers
the set of all counting numbers and zero

Numbers can be represented using digits, as points on a number line, with words, and with objects.

Example 1

Represent the number 7 in two other ways.

1. Draw a point at 7 on a number line.

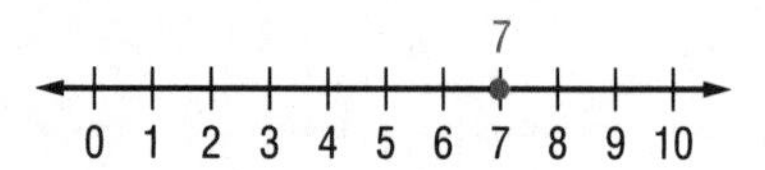

2. Write 7 with words.
seven

YOUR TURN!

Represent the number −3 in two other ways.

1. Draw a point at −3 on a number line.

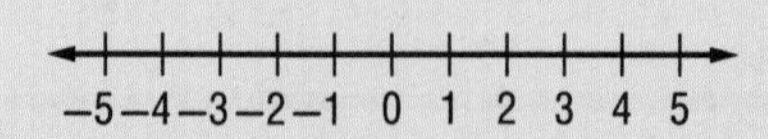

2. Write −3 with words.

____

Example 2

Circle all sets in which the number −13 belongs.

counting numbers   whole numbers   integers

1. Counting numbers are {1, 2, 3 ...}.

   −13 is not a counting number.

2. Whole numbers are {0, 1, 2, 3 ...}.

   −13 is not a whole number.

3. Integers are {..., −3, −2, −1, 0, 1, 2, 3 ...}.

   −13 is an integer.

4. Circle integers.

YOUR TURN!

Circle all sets in which the number 41 belongs.

counting numbers   whole numbers   integers

1. Counting numbers are {________}.

   41 ______ a counting number.

2. Whole numbers are {________}.

   41 ______ a whole number.

3. Integers are {________

   ________}.

   41 ______ an integer.

4. Circle ________

   ________.

Example 3

Name the opposite of −7.

1. Use a number line. Begin at zero and move left 7 units. Plot a point at −7.

2. Count the same number of units to the right of zero. Plot and label the point.

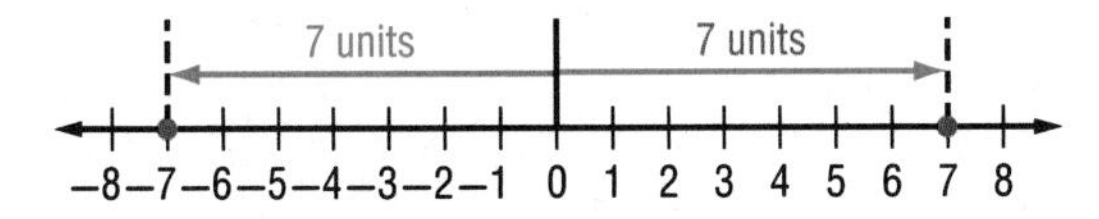

3. The opposite of −7 is 7.

YOUR TURN!

Name the opposite of 14.

1. Use a number line. Begin at zero and move ______ units. Plot a point at ______.

2. Count the same number of units to the ______ of zero. Plot and label the point.

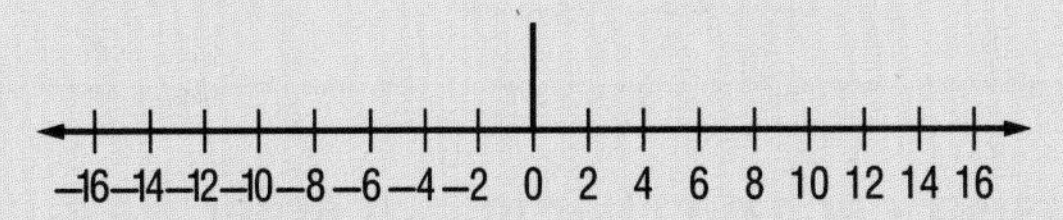

3. The opposite of 14 is ______.

GO ON

Guided Practice

Circle all sets in which each number belongs.

1. 16

counting numbers | whole numbers | integers

2. 0

counting numbers | whole numbers | integers

3. −12

counting numbers | whole numbers | integers

4. 1

counting numbers | whole numbers | integers

Name the opposite of each number.

5. 5 ______

5 units

−8 −7 −6 −5 −4 −3 −2 −1 0 1 2 3 4 5 6 7 8

6. −14 ______

−16 −14 −12 −10 −8 −6 −4 −2 0 2 4 6 8 10 12 14 16

7. −2 ______

8. 1 ______

9. 15 ______

10. −8 ______

Step by Step Practice

11. Represent the number 12 in two other ways.

Step 1 Draw a point at 12 on a number line.

Step 2 Write 12 with words.

____

Represent each number in two other ways.

12. 3 ______________

−16 −14 −12 −10 −8 −6 −4 −2 0 2 4 6 8 10 12 14 16

13. 10 ______________

−16 −14 −12 −10 −8 −6 −4 −2 0 2 4 6 8 10 12 14 16

Step by Step Problem-Solving Practice

Solve.

14 Helen got into an elevator on the ground floor and rode it down three floors. What set of numbers includes the number for the floor Helen went to?

ELEVATOR		
	0	ground floor
↓	−1	down 1 floor
↓	−2	down 2 floors
↓	−3	down 3 floors

Check off each step.

_______ **Understand: I underlined key words.**

_______ **Plan: To solve the problem, I will** _______________________.

_______ **Solve: The answer is** _______________________.

_______ **Check: I checked my answer by** _______________________.

Skills, Concepts, and Problem Solving

Name all sets in which each number belongs.

15 25

____

____

16 −51

____

____

17 0

____

____

18 1

____

____

19 −9

____

____

20 51

____

____

GO ON

Represent each number in two other ways.

21 −15 ____________________

−16 −14 −12 −10 −8 −6 −4 −2 0 2 4 6 8 10 12 14 16

22 0 ____________________

−8 −7 −6 −5 −4 −3 −2 −1 0 1 2 3 4 5 6 7 8

Name the opposite of each number.

23 10 ____________

24 −12 ____________

−16 −14 −12 −10 −8 −6 −4 −2 0 2 4 6 8 10 12 14 16

25 0 ____________

26 88 ____________

27 26 ____________

28 −6 ____________

29 1 ____________

30 100 ____________

Solve.

31 **BASKETBALL** In a basketball game, goals are worth 1, 2, and 3 points. What types of numbers are shown on the scoreboard?

____

32 **WEATHER** The temperature at 7:00 P.M. was 4°C. At 12:00 P.M. it reached the opposite temperature. What was the temperature at 12:00 P.M.?

____

Vocabulary Check **Write the vocabulary word that completes each sentence.**

33 ____________________ are numbers used to count objects.

34 Whole numbers and their opposites is the set of ____________________.

35 ____________________ are the set of all counting numbers and zero.

36 **Reflect** Name the only integer that does not have an opposite. Explain your answer.

____

____

STOP

Lesson 1-2

Compare and Order Numbers

KEY Concept

On a number line, the greater number is to the right.

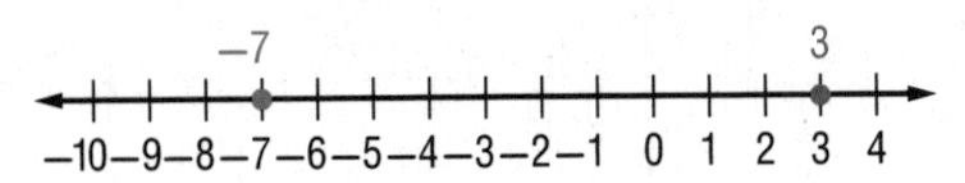

–7 is less than 3.

$-7 < 3$

Notice that the inequality symbol points to the lesser number.

3 is greater than –7.

$3 > -7$

When comparing numbers using **place value**, begin at the leftmost position and compare the digits.

$143 < 192$

In the hundreds place, 1 = 1. In the tens place, 4 < 9.

VOCABULARY

inequality
a number sentence that compares two unequal expressions and uses <, >, ≤, ≥, or ≠

negative number
a number less than zero

place value
a value given to a digit by its position in a number

positive number
a number greater than zero

When comparing negative numbers, the number closest to zero is greater.

Example 1

Use >, <, or = to compare –6 and –5.

1. Graph both numbers on a number line.

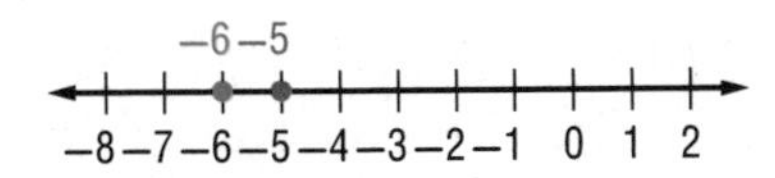

2. Because –5 is to the right of –6, it is the greater number.
3. Write an inequality.

$-6 < -5$

Point the symbol to the lesser number.

YOUR TURN!

Use >, <, or = to compare 3 and –1.

1. Graph both numbers on a number line.

2. Because ________ is to the ________ of –1, it is the ________ number.
3. Write an inequality.

3 ◯ –1

GO ON

Example 2

Use >, <, or = to compare 2,145 and 2,134.

1. Begin on the left. The digits in the thousands place and hundreds place are the same.
2. So, compare the digits in the tens place.

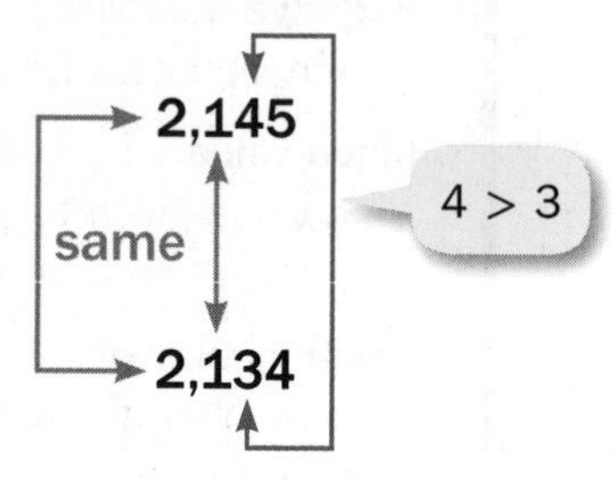

3. Write an inequality statement.

2,145 > 2,134

YOUR TURN!

Use >, <, or = to compare 11,099 and 11,809.

1. Begin on the left. The digits in the ______ place and ______ place are the ______.
2. So, compare digits in the ______ place.

11,099

11,809

3. Write an inequality statement.

11,099 ◯ 11,809

Example 3

Order 212, 202, −200, 102, and −122 from least to greatest.

1. The number farthest left is the least. The negative numbers are −200 and −122. −200 is farther to the left than −122.

   −200 < −122

2. The number closest to zero is the least positive. The positive numbers are 212, 202 and 102. 102 is closest to zero. 202 is closer to zero than 212.

   102 < 202 < 212

3. Write the numbers from least to greatest.

   −200, −122, 102, 202, 212

YOUR TURN!

Order 1,303; −2,713; −1,003; 1,297; and −2,987 from least to greatest.

1. The number farthest left is the least. The negative numbers are −2,713; −1,003; and −2,987. ______ is farther to the left than ______ and ______.

   ______ < ______ < ______

2. The number closest to zero is the least positive. The positive numbers are 1,303 and 1,297.

   ______ is closer to zero than ______.

   ______ < ______

3. Write the numbers from least to greatest.

   ______

Guided Practice

Use >, <, or = to compare each pair of numbers.

1. −19 ◯ −29

−30 −29 −28 −27 −26 −25 −24 −23 −22 −21 −20 −19 −18

2. 54 ◯ 41

40 41 42 43 44 45 46 47 48 49 50 51 52 53 54 55 56 57 58 59 60

3. 542 ◯ 533

4. 21 ◯ 211

5. −1,572 ◯ −1,610

6. 920 ◯ 902

7. −23,546 ◯ −24,563

8. 8,007 ◯ 7,008

Step by Step Practice

9. Order 78, 778, −117, 687, −68, −807 from least to greatest.

Step 1 List the negative numbers. ________________

Order the negative numbers from least to greatest. Name the number that is farthest to the left on a number line first.

____

Step 2 List the positive numbers. ________________

Order the positive numbers from least to greatest. Name the number that is closest to zero on a number line first.

____

Step 3 Write the numbers from least to greatest.

____

10. Order 304, −493, −39, 449, −94, 439 from least to greatest.

Order the negative numbers from least to greatest. ________________

Order the positive numbers from least to greatest. ________________

Write the numbers from least to greatest.

____

GO ON

Order each list from least to greatest.

11 785, −801, −818, 181, 788

____

12 12,766; −12,606; 12,677; 12,676; 12,716

____

13 146, 416, −46, −614, 14

____

14 5,003; −305; −505; −53; −3,505

____

Step by Step Problem-Solving Practice

Solve.

15 **GOLF** The six members of the Burlington Bears girls golf team received the following scores when they played a round of golf: −1, +3, +1, −2, +5, and −4. In golf, the best scores are the lowest. Order the scores from best to worst.

Order the negative numbers from least to greatest.

____

Order the positive numbers from least to greatest.

____

Write the numbers from least to greatest.

____

Check off each step.

_____ **Understand: I underlined key words.**

_____ **Plan: To solve the problem, I will** ____________________________.

____.

_____ **Solve: The answer is** ____________________________.

_____ **Check: I checked my answer by** ____________________________.

Skills, Concepts, and Problem Solving

Write each number that is described.

16 a gain of 9 pounds ______

17 6 degrees below zero ______

18 15 feet above sea level ______

19 break even ______

20 7 strokes over par ______

21 3rd floor of the basement ______

22 a loss of 2 pounds ______

23 40 feet below sea level ______

24 Write the answers from Exercises 16–21 in order from least to greatest.

______

25 Write the answers from Exercises 18–23 in order from greatest to least.

______

Solve.

26 **FOOTBALL** A series of football plays is shown.

Play	1	2	3	4	5	6	7	8
Yds gained/lost	−5	+1	−1	0	−4	+3	+2	−7

Order the plays from the most yards gained to the most yards lost.

______

27 **GOLF** The scores of the top five finishers in a local golf tournament were −5, 1, −3, 0, −2. The leader in golf earns the lowest score. List them in order from the leader to fifth place.

______

Vocabulary Check **Write the vocabulary word that completes each sentence.**

28 A(n) ______ is less than zero.

29 ______ is a value given to a digit by its position in a number.

30 **Reflect** If zero is the least integer in a list of integers, what can you conclude about the other integers in the list?

______

______

Chapter 1

Progress Check 1 (Lessons 1-1 and 1-2)

Name all sets in which each number belongs.

1. -5 ______________________ ______________________

2. 10 ______________________ ______________________

3. -22 ______________________ ______________________

4. 0 ______________________ ______________________

5. Represent the number -11 in two other ways.

____

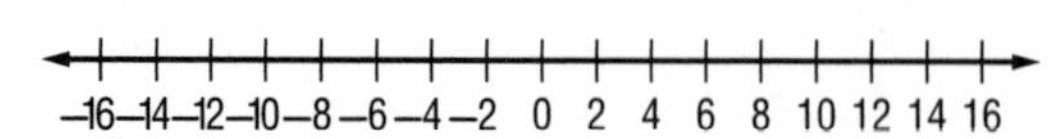

Name the opposite of each number.

6. 4 ________ 7. -16 ________ 8. -1 ________ 9. 200 ________

Use $<$, $>$, or $=$ to compare each pair of numbers.

10. 35 ◯ 53
11. 515 ◯ 505
12. 2,961 ◯ 2,691
13. 555 ◯ 777
14. 345 ◯ 234
15. 6,870 ◯ 6,780

Order each list from least to greatest.

16. 606, -606, 677, 676, -716

____

17. 1,204; $-1,242$; $-1,004$; 1,444

____

Solve.

18. **WEATHER** The high temperatures in Kevin's town last week were 0°F, 12°F, -3°F, 21°F, 15°F, -8°F, and 20°F. What were the coldest and warmest temperatures in Kevin's town last week?

____

19. **BOOKS** What number type is used when printing page numbers of a novel?

____

Lesson 1-3

Add Integers

KEY Concept

There are different methods to add integers. You can model each addend using algebra tiles.

$4 \quad + \quad (-6)$

Yellow means positive.

Red means negative.

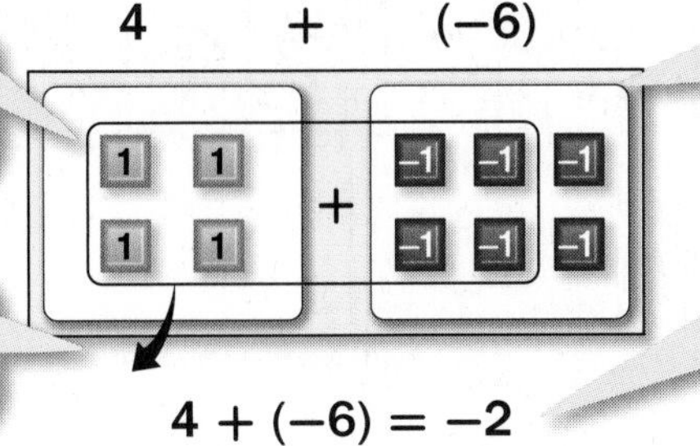

Zero pairs can be removed.

Number and color of remaining tiles name the sum.

$4 + (-6) = -2$

A number line visually shows addition of integers.

$-3 + (-5) = -8$

–9 –8 –7 –6 –5 –4 –3 –2 –1 0 1

Adding Numbers with Same Signs

When addends have the same signs, the sum has the sign of the addends.

$19 + 27 = 46 \qquad -15 + (-8) = -23$

Adding Numbers with Different Signs

When addends have different signs, follow these steps.

$-32 + 18$

1. Find the absolute value of each addend.

   $|-32| = 32 \qquad |18| = 18$

2. Subtract the absolute values of the addends.

   $32 - 18 = 14$

3. The sum has the same sign as the number with the greater absolute value.

   $-32 + 18 = -14$

–32 has the greater absolute value, so the answer is negative.

VOCABULARY

absolute value
the distance between a number and zero on a number line

addend
numbers or quantities being added together

sum
the answer to an addition problem

zero pair
two numbers when added have a sum of zero

GO ON

Example 1

Use a model to find $-8 + 5$.

1. Model each integer. Positive tiles are yellow. Negative tiles are red.
2. Form zero pairs. A zero pair is one red tile and one yellow tile.

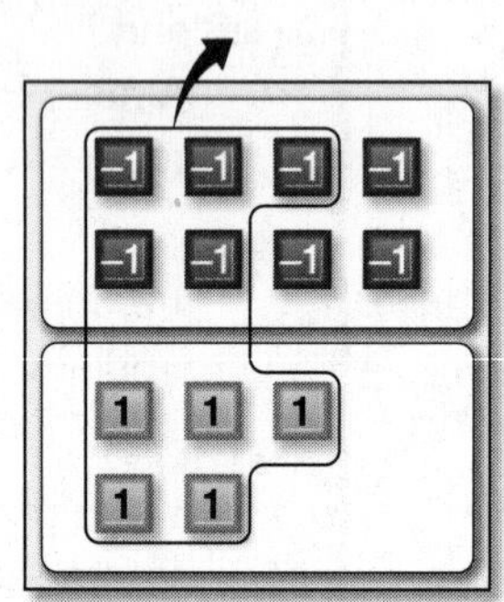

3. Count remaining tiles. There are 3 red negative tiles left.

$$-8 + 5 = -3$$

YOUR TURN!

Use a model to find $-6 + 11$.

1. Model each integer. Positive tiles are yellow. Negative tiles are red.
2. Form zero pairs. A zero pair is one red tile and one yellow tile.

3. Count remaining tiles.

There are ____________ tiles left.

$-6 + 11 =$ ____

Example 2

Use a number line to find $-3 + (-4)$.

1. From 0 on the number line, move left to -3.

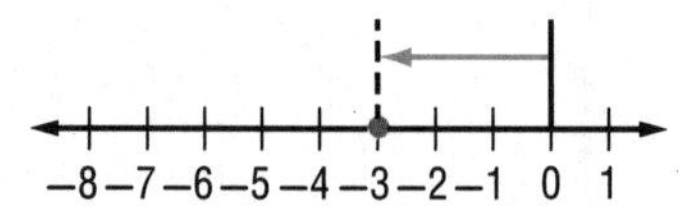

2. From -3, move left 4 units.

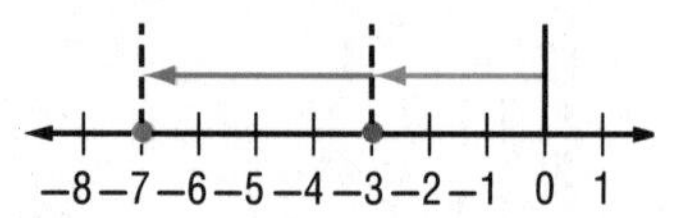

3. You ended at -7. The sum is -7.

$$-3 + (-4) = -7$$

YOUR TURN!

Use a number line to find $-6 + (-2)$.

1. From 0 on the number line, move ______ to ______.

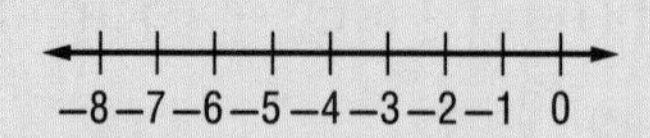

2. From ______, move ________ units.

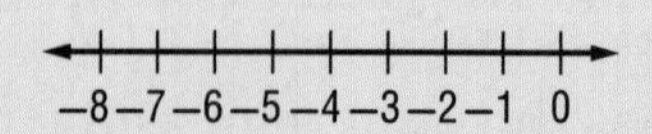

3. You ended at ______. The sum is ______.

$-6 + (-2) =$ ______

Example 3

Find 16 + (−25).

1. Are the signs of the addends the same or different?

   different

2. Find the absolute value of each addend.

   $|16| = 16$ $\quad\quad |-25| = 25$

3. What is the sign of the addend with the greatest absolute value? −

4. Subtract the absolute values of the numbers.

   $25 - 16 = 9$

5. The sum has the same sign as −25.

   $16 + (-25) = -9$

YOUR TURN!

Find 54 + (−29).

1. Are the signs of the addends the same or different?

   ______

2. Find the absolute value of each addend.

   $|54| =$ ______ $\quad\quad |-29| =$ ______

3. What is the sign of the addend with the greatest absolute value? ______

4. Subtract the absolute values of the numbers.

   ______ − ______ = ______

5. The sum has the same sign as ______.

   $54 + (-29) =$ ______

Guided Practice

Use a model to find each sum.

1 $-7 + (-3) =$ ______

2 $4 + (-2) =$ ______

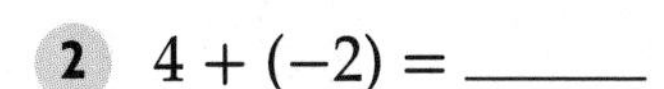

Use a number line to find each sum.

3 $-8 + -5 =$ ______

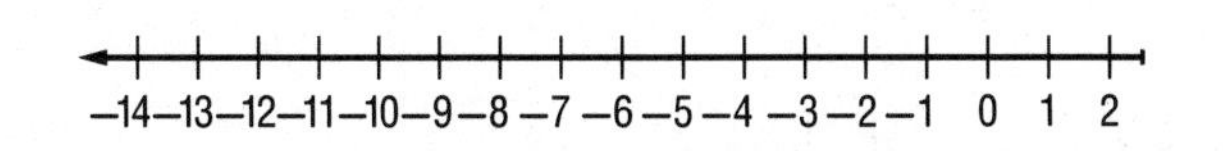

4 $-7 + (-4) =$ ______

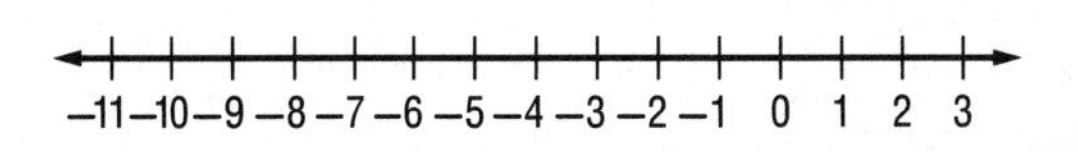

GO ON

Step by Step Practice

5 Find $-87 + 45$.

Step 1 Are the signs of the addends the same or different?

____

Step 2 Find the absolute value of each addend.

$|-87| =$ ____ $|45| =$ ____

Step 3 What is the sign of the addend with the greatest absolute value? ____

Step 4 Subtract the absolute values of the numbers.

____ − ____ = ____

Step 5 The sum has the same sign as ______.

$-87 + 45 =$ ______

Find each sum.

6 $62 + (-28) =$ ______
The signs of the addends

are ____________.

|____| = ____ |____| = ____

____ − ____ = ____

7 $15 + (-33) =$ ______
The signs of the addends

are ____________.

|____| = ____ |____| = ____

____ − ____ = ____

8 $-58 + (-99) =$ ______
The signs of the addends

are ____________.

|____| + |____|

= ____ + ____

= ____

9 $-61 + (-79) =$ ______
The signs of the addends

are ____________.

|____| + |____|

= ____ + ____

= ____

10 $45 + (-27) =$ ______

11 $-83 + (-16) =$ ______

Step by Step Problem-Solving Practice

Solve.

12 **ELEVATOR** Leon got on the elevator at the 8th floor. He rode the elevator up 10 floors, then down 15 floors, and got off the elevator. On what floor did Leon get out of the elevator?

A vertical number line can model the path of the elevator.

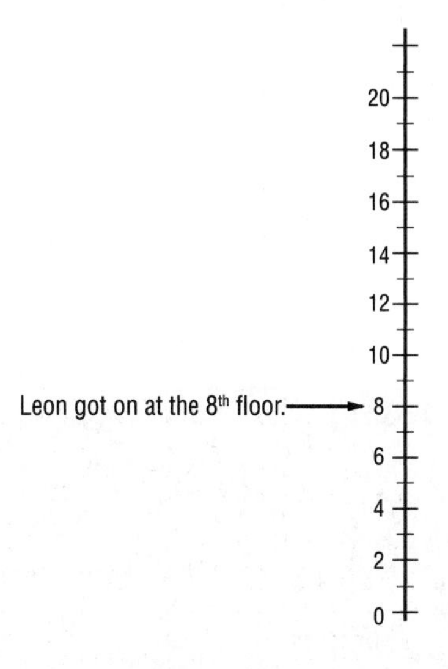

Leon got out on the ______ floor.

Check off each step.

______ **Understand: I underlined key words.**

______ **Plan: To solve the problem, I will** ______________________.

______ **Solve: The answer is** ______________________.

______ **Check: I checked my answer by** ______________________.

Skills, Concepts, and Problem Solving

Find each sum.

13 $-47 + 12 =$ ________

14 $74 + (-14) =$ ________

15 $-8 + (-11) =$ ________

16 $98 + (-36) =$ ________

17 $-75 + 86 =$ ________

18 $5 + (-22) =$ ________

19 $17 + (-88) =$ ________

20 $16 + 112 =$ ________

21 $156 + (-48) =$ ________

Find each sum.

22 $5 + 6 + (-3) =$ ________

23 $14 + (-3) + 10 =$ ________

24 $-20 + 5 + (-10) =$ ________

25 $9 + (-3) + 15 =$ ________

26 $-14 + (-8) + 25 =$ ________

27 $-4 + (-13) + (-39) =$ ________

Solve.

28 **GARDENING** Sherita plants a 5-inch tall plant in a 2-inch deep hole. The plant grows 9 inches. How tall does Sherita's plant stand above the ground?

____

29 **FOOTBALL** Ken's team starts with the football on the 20-yard line. In the next three plays they gain 8 yards, lose 2 yards, and gain 15 yards. At what yard line are they?

____

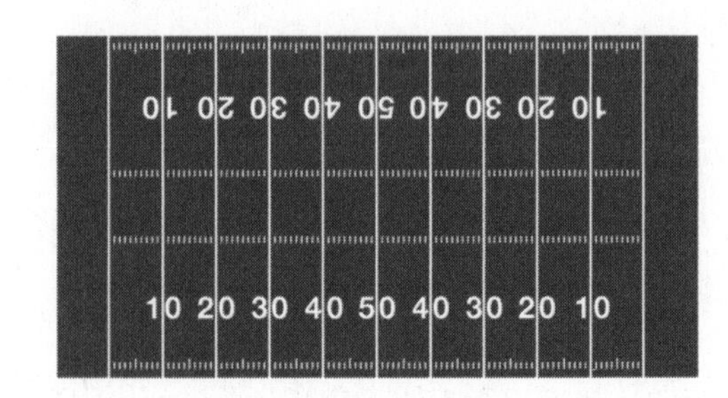

30 **MONEY** Larisa earns \$35. She spends \$16 for dinner and a movie. How much money does she have left?

____

Vocabulary Check **Write the vocabulary word that completes each sentence.**

31 Numbers or quantities added together are ________________.

32 Two numbers that have a sum of zero are called a ________________.

33 The ________________ of a number is the distance between a number and zero on a number line.

34 **Reflect** Write two different addition problems that have a negative sum, one with addends having the same sign, and one with addends having different signs. Explain why the sums are negative.

____

____

____

STOP

Lesson 1-4

Subtract Integers

KEY Concept

Subtracting an integer is the same as adding the opposite of that integer. You can change subtraction to addition by changing the sign of the second integer.

+5 is the opposite of −5.

+6 is the opposite of −6.

17 − (−5)	**−9 − (−6)**
17 + (+5)	**−9 + (+6)**
17 + 5 = 22	**−9 + 6 = −3**

−18 is the opposite of +18.

−4 is the opposite of 4.

11 − 18	**−23 − 4**
11 + (−18) = −7	**−23 + (−4) = −27**

Use the rules for adding integers.

VOCABULARY

difference
the answer to a subtraction problem

opposite
two different numbers that are the same distance from 0 on a number line

Example 1

Use a model to find −6 − 5.

1. Change to addition and use the opposite of 5.
   −6 + (−5)
2. Model each integer. There are no zero pairs.

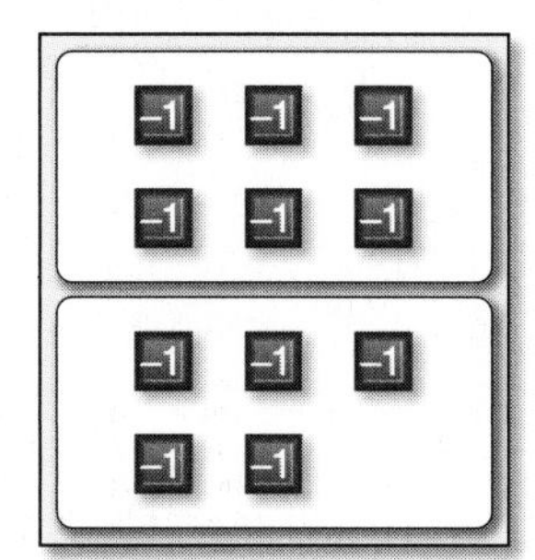

3. There are 11 red negative tiles.
   −6 − 5 = −11

YOUR TURN!

Use a model to find 4 − 9.

1. Change to addition and use the opposite of ____.
   4 ____ (____)
2. Model each integer. There are ____ zero pairs.

3. There are ______________ tiles.
   4 − 9 = ____

GO ON

Example 2

Find 42 − (−38).

1. Change to addition and use the opposite of −38.
   42 + (38)

2. The signs are the same. Add the absolute values.
   |42| + |38| = 80

3. So, 42 − (−38) = 80

YOUR TURN!

Find 8 − (−12).

1. Change to addition and use the opposite of ______.
   8 ____ (____)

2. The signs are the ______. Add the absolute values.
   |8| + |12| = ____

3. So, 8 − (−12) = ____

Example 3

Find −4 − (−10).

1. Change to addition and use the opposite of −10.
   −4 + (10)

2. The signs are different. Subtract the absolute values.
   |10| − |4| = 10 − 4 = 6

3. So, −4 − (−10) = 6

YOUR TURN!

Find −92 − (−14).

1. Change to addition and use the opposite of ______.
   −92 + (____)

2. The signs are ________. Subtract the absolute values.
   |−92| − |14| = ____ − ____ = ____

3. So, −92 − (−14) = ______

Guided Practice

Use a model to find each difference.

1. −9 − (−6) = ____

2. 11 − (−4) = ____

Find each difference.

3. $-4 - 15 =$ ______

Change to addition. $-4 + ($______$)$

The signs are ____________, so ____________ the absolute values.

$|-4|$ ____ $|-15| =$ ______

4. $-12 - (-19) =$ ______

Change to addition. $-12 + ($______$)$

The signs are ____________, so ____________ the absolute values.

$|19|$ ____ $|-12| =$ ______

Step by Step Practice

5. Find $-47 - (-19)$.

Step 1 Change to addition and use the opposite of -19. $-47 + 19$

Step 2 The signs are different. Subtract the absolute values.

______ $-$ ______ $=$ ______

Step 3 $-47 - (-19) =$ ______

Find each difference.

6. $-55 - (-18) =$ ______

Change to addition. $-55 + ($____$)$

The signs are __________, so __________ the absolute values.

$|-55|$ ____ $|18| =$ ____

7. $-24 - 86 =$ ______

Change to addition. $-24 + ($_____$)$

The signs are __________, so _____ the absolute values.

$|-24|$ ____ $|-86| =$ ______

GO ON

Step by Step Problem-Solving Practice

Solve.

8 WEATHER At 6 P.M. the temperature was 12°C below zero. When Ravi checked at 10 P.M. the temperature had fallen 6°. What was the temperature when Ravi checked?

Starting temperature	−	Degrees of change	=	New temperature
	−		=	

$\_\_\_\_ - \_\_\_\_ = x$

$\_\_\_\_ + (\_\_\_\_) = x$

$\_\_\_\_ = x$

The temperature at 10 P.M. was ________.

Check off each step.

_____ Understand: I underlined key words.

_____ Plan: To solve the problem, I will ______________________.

_____ Solve: The answer is ______________________.

_____ Check: I checked my answer by ______________________.

Skills, Concepts, and Problem Solving

Find each difference.

9 $-7 - (-4) =$ _____

10 $4 - 10 =$ _____

11 $-5 - (-5) =$ _____

12 $-17 - 23 =$ _____

13 $68 - (-11) =$ _____

14 $-38 - (-4) =$ _____

15 $41 - (-33) =$ _____

16 $19 - 200 =$ _____

17 $-106 - (-18) =$ _____

Find each difference.

18 $99 - (-13) =$ ______

19 $-72 - (-72) =$ ______

20 $-54 - 23 =$ ______

21 $93 - (-13) =$ ______

22 $89 - 350 =$ ______

23 $6 - 48 =$ ______

24 $3 - 103 =$ ______

25 $90 - (-90) =$ ______

26 $-24 - 24 =$ ______

Solve.

27 **WEATHER** The temperature where David lives was −4°F. When he talked to his friend Silvia, she read the temperature from the thermometer shown at the right. What was the difference in temperatures from where David lives to where Silvia lives?

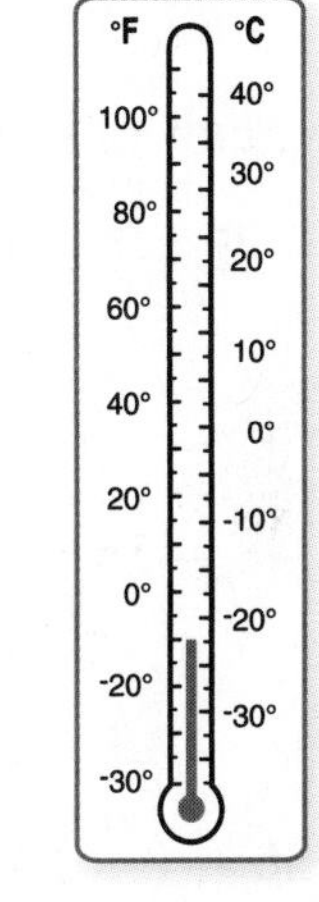

____

28 **SCUBA DIVING** Carlie descended 45 feet on her first dive. For her second dive, she descended 78 feet. What was the difference in the depth of Carlie's first dive to her second dive?

____

29 **PARKING GARAGE** Felicia goes to a doctor's office that has a parking garage below the building. She parks her car four levels below the street level. Her doctor's office is located on the fifth floor. When she leaves the doctor's office to return to her car, how many levels does she ride the elevator?

____

Vocabulary Check **Write the vocabulary word that completes each sentence.**

30 Two numbers, that are the same distance from 0 on a number line are ______________.

31 The ______________ is the answer to a subtraction problem.

32 **Reflect** Write a subtraction problem that includes two negative integers, but has a difference that is a positive integer.

____

____

STOP

Chapter 1

Progress Check 2 (Lessons 1-3 and 1-4)

Find each sum.

1. $38 + (-13) =$ _____
2. $-65 + (73) =$ _____
3. $-22 + (-55) =$ _____
4. $51 + 35 =$ _____
5. $73 + (-11) =$ _____
6. $-1 + (-19) =$ _____
7. $117 + (-78) =$ _____
8. $-5 + 68 =$ _____
9. $235 + (-99) =$ _____

Find each difference.

10. $-25 - (-3) =$ _____
11. $16 - (96) =$ _____
12. $-4 - (-102) =$ _____
13. $-8 - 15 =$ _____
14. $77 - (-4) =$ _____
15. $-3 - 40 =$ _____
16. $36 - (-8) =$ _____
17. $350 - (-350) =$ _____
18. $-29 - 18 =$ _____

Solve.

19. **MONEY** Enrico had $35 in his pocket. He then found a hole in his pocket and the money was missing. He also owes his mother $12. What number represents the amount Enrico has?

____

20. **CHARITY** When Tamera puts her hair into a ponytail, it is 19 inches long. In order to donate her hair to *Locks of Love*, her ponytail needs to be at least 12 inches. Write an equation using a negative integer which shows the length of Tamera's hair after her donation.

____

Lesson 1-5

Multiply Integers

KEY Concept

Multiplication is repeated addition. The expression $3 \cdot 5$ means 3 groups of 5 or $5 + 5 + 5 = 15$.

To multiply integers, you can model each **factor** using algebra tiles. For $2 \cdot (-3)$, use 2 sets of 3 negative tiles.

2 is the number of rows. 3 is the number in each row.

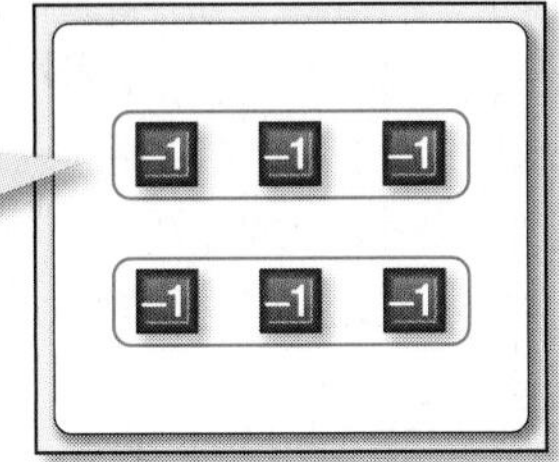

$2 \cdot (-3) = -6$

The product of –6 is the total number of negative tiles.

To multiply integers without a model, multiply their **absolute values**, then write the sign of the product.

If the integers are the same sign, the product is positive.

$7 \cdot 6 = 42$ **$-8 \cdot (-5) = 40$**

If the integers are different signs, the product is negative.

$4 \cdot (-3) = -12$ **$-9 \cdot 10 = -90$**

VOCABULARY

absolute value
the distance between a number and zero on a number line

factor
a number that divides into a whole number evenly; also a number that is multiplied by another number

multiplication
an operation on two numbers to find their product

product
the answer to a multiplication problem

zero pair
two numbers when added that have a sum of zero

The order in which the factors are multiplied does not change the product, so $-5 \cdot 8 = 8 \cdot (-5)$.

Example 1

Find $-3 \cdot 4$.

1. Place the positive integer first.
   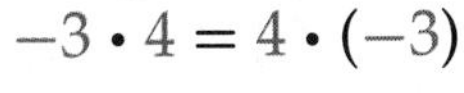
   $-3 \cdot 4 = 4 \cdot (-3)$
2. Model using 4 groups of -3.
3. Count the tiles.
   There are 12 red negative tiles.
   $-3 \cdot 4 = -12$

YOUR TURN!

Find $-5 \cdot 2$.

1. Place the positive integer first.
   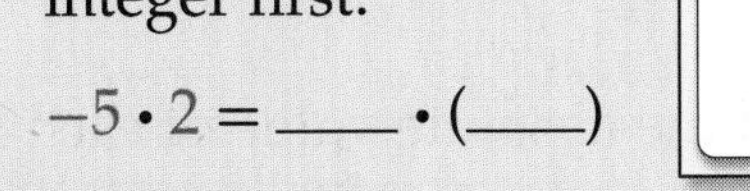
   $-5 \cdot 2 =$ ____ $\cdot$ (____)
2. Model using
   ____ groups of ____
3. Count the tiles.
   There are ________ negative tiles.
   $-5 \cdot 2 =$ ____

GO ON

Example 2

Find $-12 \cdot 87$.

1. Find the absolute value of each factor.
   $|-12| = 12$ $|87| = 87$

2. Write the problem in vertical form. Multiply.

$$\begin{array}{r} 87 \\ \times\ 12 \\ \hline 174 \\ +\ 870 \\ \hline 1{,}044 \end{array}$$

3. The product of two factors with different signs is negative.

   $-12 \cdot 87 = -1{,}044$

YOUR TURN!

Find $-23 \cdot (-36)$.

1. Find the absolute value of each factor.

   $|-23| =$ ____ $|-36| =$ ____

2. Write the problem in vertical form. Multiply.

$$\begin{array}{r} 36 \\ \times\ 23 \\ \hline \\ +\ \underline{\quad\quad} \end{array}$$

3. The product of 2 factors with __________ signs is __________.

   $-23 \cdot (-36) =$ ______

Guided Practice

Find each product.

1. $-3 \cdot 3 =$ ______

2. $-6 \cdot 3 =$ ______

Step by Step Practice

3. Find $-14 \cdot 26$.

$$\begin{array}{r} 26 \\ \times\ 14 \\ \hline \\ +\ \underline{\quad\quad} \end{array}$$

Step 1 Find the absolute value of each factor.

Step 2 Write the problem in vertical form. Multiply.

Step 3 The product of two factors with __________ signs is

__________.

$-14 \cdot 26 =$ ________

Find each product.

4 $-13 \cdot (-9) =$ ______

$$\begin{array}{r} 13 \\ \times\ 9 \\ \hline \end{array}$$

The product of two factors with ______ signs is ______.

5 $15 \cdot (-7) =$ ______

$$\begin{array}{r} 15 \\ \times\ 7 \\ \hline \end{array}$$

The product of two factors with ______ signs is ______.

6 $-37 \cdot 12 =$ ______

$$\begin{array}{r} 37 \\ \times\ 12 \\ \hline \end{array}$$

7 $-86 \cdot (-11) =$ ______

$$\begin{array}{r} 86 \\ \times\ 11 \\ \hline \end{array}$$

Step by Step Problem-Solving Practice

Solve.

8 Omar has $225 saved from his birthday. He needs to decide if he has enough money to buy an MP3 player. If the store allows him to pay $18 per month over the next year, does he have enough money? Show your work.

Saved amt.	+	# of mo.	•	Amt. owed ea. mo.	=	Final amt. saved
	+		•		=	*a*

______ + ______ • (______) = a

______ + (______) = a

______ = a

Check off each step.

______ **Understand: I underlined key words.**

______ **Plan: To solve the problem, I will** ______________________.

______ **Solve: The answer is** ______________________.

______ **Check: I checked my answer by** ______________________.

GO ON

Skills, Concepts, and Problem Solving

Find each product.

9. $-8 \cdot (-6) =$ ________
10. $-3 \cdot (-24) =$ ________
11. $27 \cdot (-11) =$ ________
12. $-46 \cdot 23 =$ ________
13. $-56 \cdot 7 =$ ________
14. $-15 \cdot 7 =$ ________
15. $3 \cdot (-4) =$ ________
16. $-81 \cdot (-9) =$ ________
17. $14 \cdot (-93) =$ ________

Solve.

18. **GAME SHOW** Kira has a score of 0 points. If she answers three 15-point questions incorrectly, what is her new score?

____

19. **OCEAN** A submarine descends into the ocean 9 feet per second. What is the depth of the submarine after 1 minute?

____

20. **WEATHER** The temperature is 15°F at 5:00 A.M. and the temperature drops 3°F every hour. What will the temperature be at 11:00 A.M.?

____

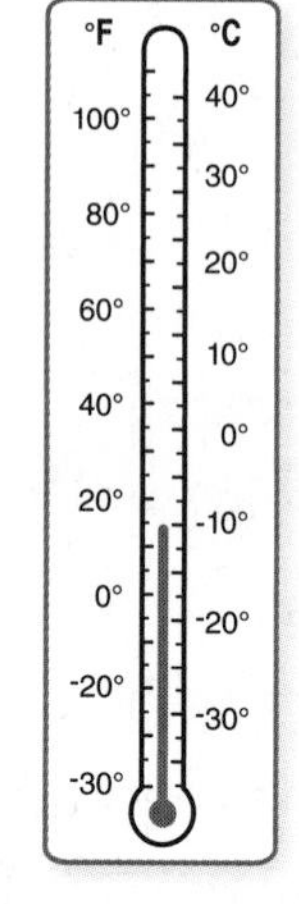

Vocabulary Check **Write the vocabulary word that completes each sentence.**

21. The answer to a multiplication problem is a(n) _______________.
22. A number multiplied by another number is a(n) _______________.
23. A(n) _______________ is two numbers when added have a sum of zero.
24. **Reflect** When three negative factors are multiplied, is the product positive or negative? Explain.

____

____

____

____

STOP

Lesson 1-6

Divide Integers

KEY Concept

To divide integers using algebra tiles, separate the tiles into equal groups. For $-8 \div (2)$, use 8 negative tiles.

The 8 red tiles is the dividend, −8.

The divisor is 2, or the number of equal groups.

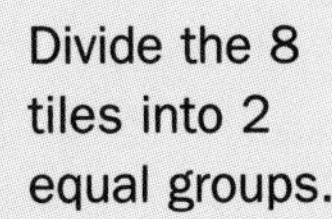

Count the tiles in each group. This is the quotient.

$$8 \div (2) = -4$$

Division is the **inverse operation** of multiplication. Use multiplication facts when you divide.

$$12 \div (-3) = -4$$

$$-3 \cdot -4 = 12$$

The sign rules are the same as multiplication.

If the integers are the same sign, the quotient is positive.

If the integers are different signs, the quotient is negative.

VOCABULARY

dividend
a number that is being divided

divisor
the number by which the dividend is being divided

inverse operations
operations that undo each other

quotient
the answer to a division problem

To divide integers without a model, find their absolute values. Then use the sign rules and write the sign of the quotient.

Example 1

Find $10 \div 2$.

1. Model 10 using yellow positive tiles.
2. Arrange tiles into 2 equal groups.
3. There are 2 groups of 5.
4. Count the tiles in one of the groups.

$10 \div 2 = 5$

YOUR TURN!

Find $-9 \div 3$.

1. Model _____ using _____________ tiles.
2. Arrange tiles into _____ equal groups.
3. There are _____ groups of _____.
4. Count the tiles in one of the groups.

$-9 \div 3 =$ _____

GO ON

Example 2

Find $-208 \div (-16)$.

1. Find the absolute value of each number.

   $|-208| = 208$　　$|-16| = 16$

2. Write the problem in vertical form. Divide.

$$\begin{array}{r} 13 \\ 16\overline{)208} \\ -16 \\ \hline 48 \\ -48 \\ \hline 0 \end{array}$$

3. The quotient of two factors with the same signs is positive. The quotient is 13.

   $-208 \div (-16) = 13$

YOUR TURN!

Find $-187 \div 11$.

1. Find the absolute value of each number.

   $|-187| =$ ______　　$|11| =$ ______

2. Write the problem in vertical form. Divide.

$$\begin{array}{r} 11\overline{)187} \\ -\underline{\quad} \\ \\ -\underline{\quad} \end{array}$$

3. The quotient of two factors with ______________ is ______________.

   The quotient is ______.

   $-187 \div 11 =$ ______

Guided Practice

Find each quotient.

1. $-15 \div 3 =$ ______

2. $8 \div 4 =$ ______

Step by Step Practice

3. Find $-270 \div (-18)$.

Step 1 Find the absolute value of each number.

$|-270| =$ ______　　$|-18| =$ ______

Step 2 Write the problem in vertical form. Divide.

$$\begin{array}{r} 18\overline{)270} \\ \underline{\quad\quad} \\ \\ \underline{\quad\quad} \end{array}$$

Step 3 Both the dividend and divisor have the ______________.

The quotient is ______________.　The quotient is ______.

Find each quotient.

4 $-722 \div 19 =$ ______

19)722
−
———
−
———

5 $-242 \div (-11) =$ ______

11)242
−
———
−
———

6 $312 \div (-13) =$ ______

13)312

7 $-456 \div 19 =$ ______

19)456

Step by Step Problem-Solving Practice

Solve.

8 Experts believe 100,000 cheetahs were living worldwide 100 years ago. Today, they believe only 10,000 cheetahs exist. If the yearly decrease was the same, what was the change in the cheetah population for each of the last 100 years?

The difference in populations	divided by	100 years	finds	the average change per year.
(−)	÷		=	
	÷		=	

Each year the cheetah population changed by an average of ______________.

Check off each step.

______ Understand: I underlined key words.

______ Plan: To solve the problem, I will ______________.

______ Solve: The answer is ______________.

______ Check: I checked my answer by ______________.

GO ON

Skills, Concepts, and Problem Solving

Find each quotient.

9. $-12 \div 3 =$ _____

10. $-10 \div 2 =$ _____

11. $-40 \div 5 =$ _____

12. $64 \div (-8) =$ _____

13. $-56 \div (-7) =$ _____

14. $-608 \div 8 =$ _____

15. $189 \div (-27) =$ _____

16. $456 \div 24 =$ _____

Solve.

17. **FINANCE** The stock market plunged 375 points over 3 days. If the stocks fell by the same amount each day, how much was the daily fall?

____

18. **BASEBALL** The attendance at the Pittsburgh Pirates'games has decreased from 2001 to 2007. Use the table to find the mean decrease in average game attendance.

____

Year	2001	2007
Average Game Attendance	30,742	22,139

Vocabulary Check **Write the vocabulary word that completes each sentence.**

19. Operations that undo each other are _____________.

20. The _____________ is the number by which the dividend is divided.

21. A number that is being divided is the _____________.

22. **Reflect** How are the rules for multiplying integers and dividing integers similar?

____

____

____

STOP

Chapter 3 Progress Check 3 (Lessons 1-5 and 1-6)

Find each product.

1 $-4 \cdot (-62) =$ ______

2 $33 \cdot (-5) =$ ______

3 $(-17) \cdot 11 =$ ______

4 $-7 \cdot (-12) =$ ______

5 $13 \cdot (-14) =$ ______

6 $-68 \cdot 15 =$ ______

7 $81 \cdot 5 =$ ______

8 $-2 \cdot 58 =$ ______

9 $185 \cdot 8 =$ ______

10 $-39 \cdot 12 =$ ______

11 $-9 \cdot (-116) =$ ______

12 $32 \cdot (-13) =$ ______

Find each quotient.

13 $-490 \div 14 =$ ______

14 $-1{,}136 \div (-16) =$ ______

15 $784 \div (-28) =$ ______

16 $-28 \div (-14) =$ ______

17 $39 \div (-13) =$ ______

18 $-896 \div 28 =$ ______

19 $96 \div 6 =$ ______

20 $322 \div (-7) =$ ______

21 $-1{,}482 \div 19 =$ ______

22 $-1{,}131 \div 13 =$ ______

23 $-180 \div (-12) =$ ______

24 $416 \div (-8) =$ ______

Solve.

25 **CARD GAME** In a game where red cards deduct 3 points from a player's score, Romeo drew seven red cards. What was the change in Romeo's score?

______

26 **SHOPPING** The price on an item dropped $145 during a 5-day sale. The price changed the same each day. What was the daily change in price?

______

Lesson 1-7

Solve One-Step Equations

KEY Concept

The solution of a one-step equation names the value of the **variable** that makes the equation true. One way to solve is to model the equation using algebra tiles.

Make equal groups of algebra tiles to solve this equation.

$2x = -6$

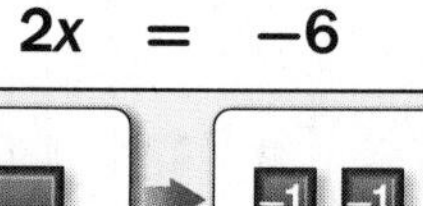

The coefficient of x tells you how many equal groups to make.

$x = -3$

Make zero pairs to solve this equation.

$x + 4 = 9$

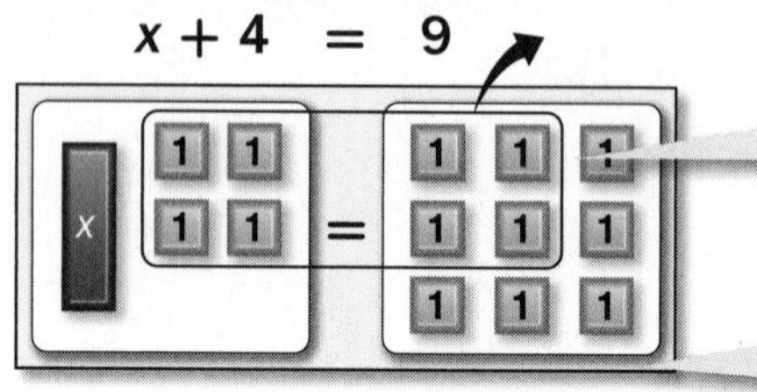

Remove the same number of tiles from each side.

The number remaining on the right side is the value of x.

$x = 5$

You can also use **inverse operations** to isolate the variable on one side of the equal sign.

VOCABULARY

inverse operations
operations that undo each other

one-step equations
an equation that can be solved in one step

variable
a letter or symbol used to represent an unknown quantity

zero pair
two numbers when added have a sum of zero

Example 1

Solve the equation $3x = 6$.

1. Model $3x = 6$.

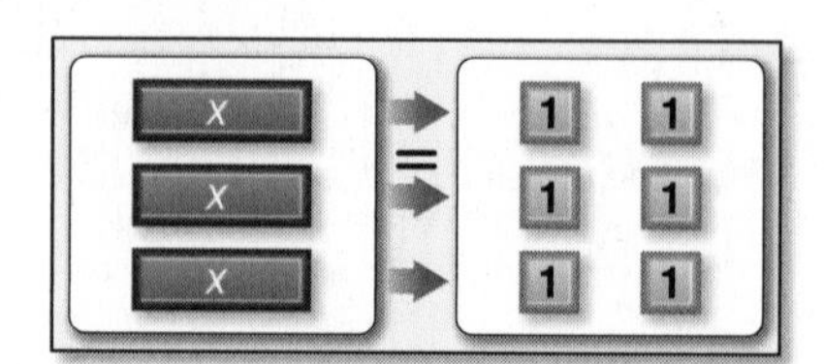

2. There are 3 x-tiles. Arrange the tiles into 3 equal groups.

3. Count the tiles paired with one x-tile.
$x = 2$

YOUR TURN!

Solve the equation $2x = -8$.

1. Model $2x = -8$.

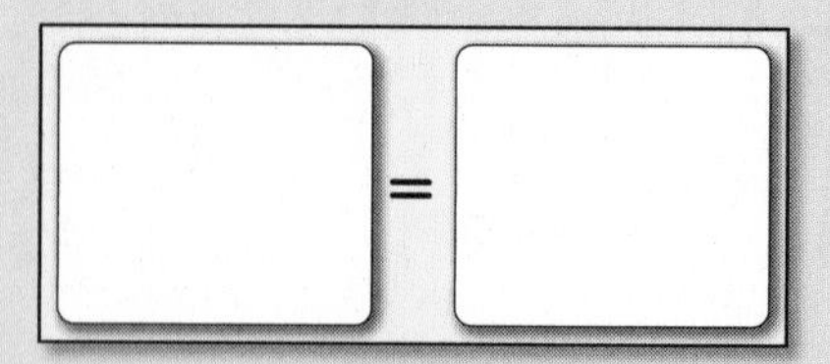

2. There are _____ x-tiles. Arrange the tiles into _____ equal groups.

3. Count the tiles paired with one x-tile.
$x =$ _____

Example 2

Solve the equation $x - 3 = -2$.

1. Model $x - 3 = -2$.

$x - 3 = 2$ is the same as $x + (-3) = 2$

2. To isolate the x-tile, add 3 positive tiles to the left mat to make 3 zero pairs. Add 3 positive tiles to the right side.

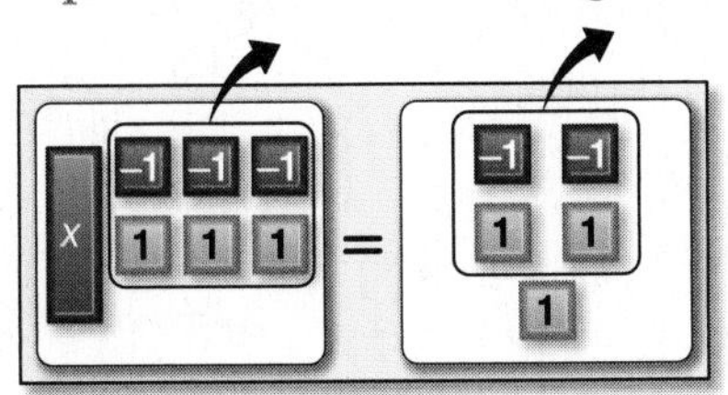

3. There are 3 zero pairs on the left side and 2 zero pairs on the right side. Remove them.

4. Count the number of tiles on the right mat.

$x = 1$

YOUR TURN!

Solve the equation $x - 2 = 5$.

1. Model $x - 2 = 5$.

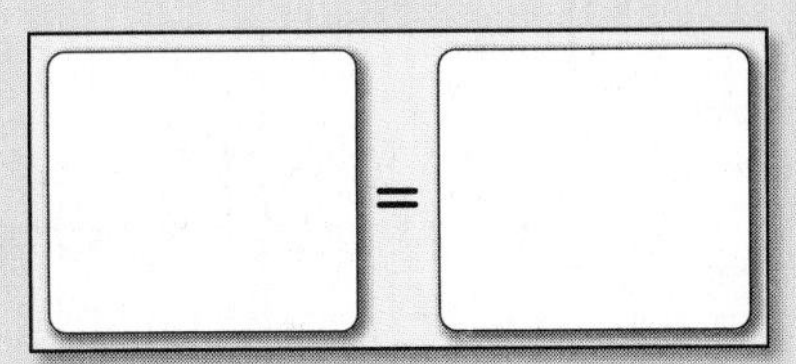

2. To isolate the x-tile, add ________ tiles to the left mat to make ____ zero pairs. Add ________ tiles to the right side.

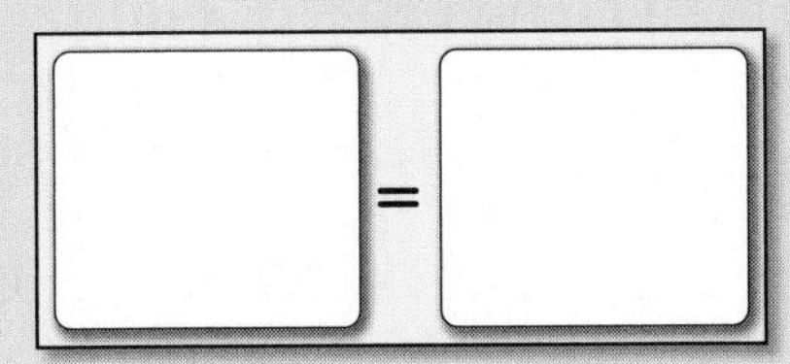

3. Remove ____ zero pairs from the ______ side.

4. Count the tiles on the right side of the mat.

$x =$ ____

Example 3

Solve the equation $\frac{x}{6} = 4$.

1. The inverse operation of division is multiplication.

2. Multiply each side of the equation by 6.

$$\not{6}\left(\frac{x}{\not{6}}\right) = (4)6$$

$$x = 24$$

YOUR TURN!

Solve the equation $\frac{x}{3} = 13$.

1. The inverse operation of division is __________.

2. Multiply each side of the equation by ____.

$$\_\_\_\left(\frac{x}{3}\right) = (13)\_\_\_$$

$$x = \_\_\_$$

GO ON

Guided Practice

Solve each equation.

1 $3x = -12$

$x =$ ______

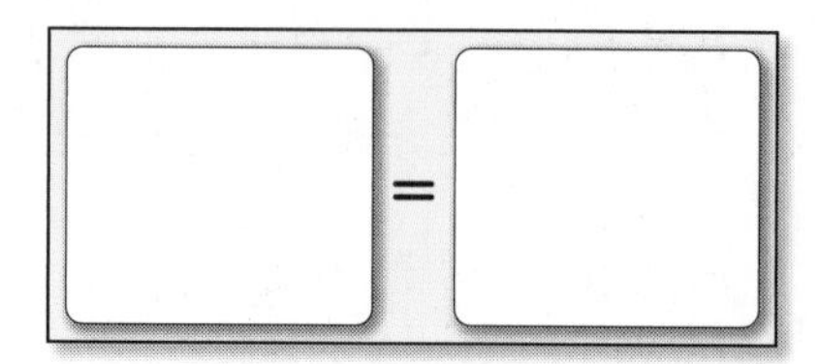

2 $2x = 2$

$x =$ ______

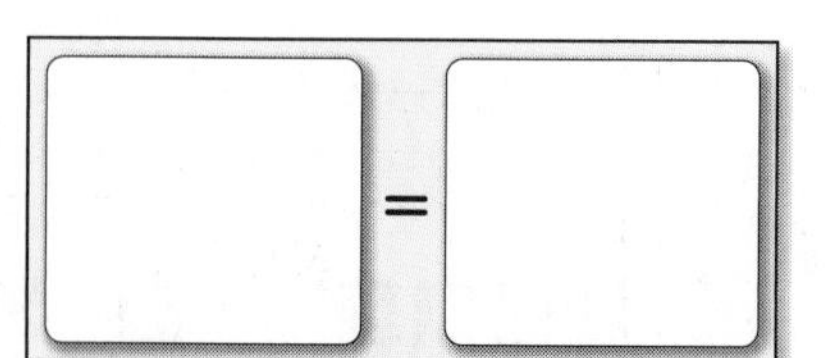

Step by Step Practice

3 Solve the equation $y - 12 = -3$.

Step 1 The inverse operation of subtraction is ____________.

Step 2 Add ______ to each side of the equation.

$y - 12 = -3$

______ ______

$y =$ ______

$-3 + 12$
The signs are different, so subtract the absolute values.

Solve each equation.

4 $\frac{x}{3} = -9$

______ $\left(\frac{x}{3}\right) = (-9)$ ______

______ = ______

Check: $\frac{\square}{3} =$ ______

$\frac{x}{3} = \frac{\square}{3} =$ ______

5 $14x = -98$

$\frac{14x}{\square} = \frac{-98}{\square}$

$x =$ ______

Check: 14 ______ = ______

$14x = 14$ ______ = ______

6 $x - 7 = 2$

______ + ______ = ______

______ = ______

$x =$ ______

Check: ______ $- 7 = 2$

7 $x - (-2) = -5$

______ + ______ = ______

______ = ______

$x =$ ______

Check: ______ $- (-2) = -5$

Step by Step Problem-Solving Practice

Solve.

8. Vanesa is training for a marathon that takes place next month. She ran 19 miles over the weekend. If she ran 8 miles on Sunday, how far did she run on Saturday?

______ Distance ran	+	______ distance ran	=	______ distance
	+		=	

______ + ______ = ______

______ ______

______ = ______ Vanesa ran ____________ on Saturday.

Check off each step.

______ **Understand: I underlined key words.**

______ **Plan: To solve the problem, I will** ______________________________.

______ **Solve: The answer is** ______________________________.

Skills, Concepts, and Problem Solving

Solve each equation. Check your answers

9. $8c = -64$

   $c =$ ______

10. $26 = 9 + g$

    $g =$ ______

11. $b - 16 = 51$

    $b =$ ______

12. $\frac{x}{-5} = 19$

    $x =$ ______

13. $m + 15 = 9$

    $m =$ ______

14. $\frac{x}{7} = -4$

    $x =$ ______

15. $28 + a = -16$

    $a =$ ______

16. $-3k = -42$

    $k =$ ______

17. $d - 25 = -25$

    $d =$ ______

GO ON

Determine if each solution is correct. Write *yes* or *no*.

18 $\frac{w}{4} = 36$

$w = 9$

____

19 $p + (-3) = 17$

$p = 20$

____

20 $\frac{x}{-5} = 55$

$x = 11$

____

Solve.

21 **JEWELRY** Ebony is buying beads to make necklaces for some of her friends. She buys a container of 960 beads and each necklace requires 120 beads. How many necklaces will she be able to make?

____

22 **FUNDRAISER** Kentdale High School raised $4,522 for their annual Walk-a-Thon to improve their athletic fields. If the freshman class raised $1,890, how much did the rest of the students raise?

____

23 **GEOMETRY** The area of a rectangle is 108 square inches. Its length is 9 inches. Use the formula $A = \ell w$ to find the width of the rectangle.

____

Vocabulary Check **Write the vocabulary word that completes each sentence.**

24 A _______________ is a letter or symbol used to represent an unknown quantity.

25 The integers 2 and −2 are considered a _______________, because when added together they have a sum of zero.

26 **Reflect** Jonathan and Irwin are debating the answer to the equation $x - (-5) = 11$. Jonathan thinks $x = 6$ and Irwin believes $x = 16$. Who is correct? Explain.

____

____

STOP

Lesson 1-8

Graph Inequalities

KEY Concept

To see all the numbers that are a solution to an inequality, graph it on a number line.

Use these rules when graphing inequalities.

Open

The $>$ and $<$ mean "greater than" or "less than." They do not include the numbers in the inequality.

$x > 4$

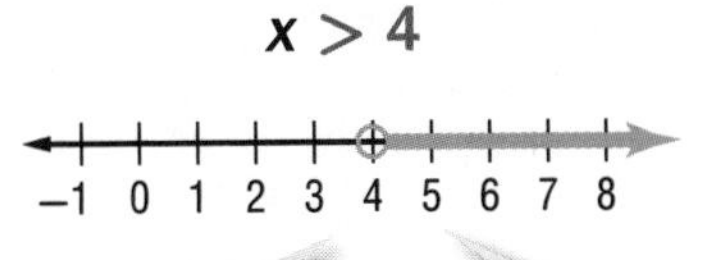

Use an open circle because 4 is not part of the solution.

Shade to the right for "greater than."

Closed

The $\geq$ and $\leq$ mean "greater than or equal to" or "less than or equal to." They do include the numbers in the inequality.

$x \leq 7$

4 5 6 7 8 9 10 11 12

Shade to the left for "less than or equal to."

Use a closed circle because 7 is part of the solution.

VOCABULARY

inequality
a number sentence that compares two unequal expressions and uses $<$ (is less than), $>$ (is greater than), $\leq$ (is less than or equal to), $\geq$ (is greater than or equal to) or $\neq$ (is not equal to)

solution
value or values of the variable makes an inequality or equation true

You can check your graph by choosing any number that is on the shaded part of the number line. Substitute that number into the inequality. If it is true, then the graph is correct.

GO ON

Example 1

Graph the inequality.
$x \leq 18$

1. Draw a number line that includes 18.
2. The point at 18 will be a closed circle. Draw the point.
3. The inequality includes numbers less than 18, so shade the line to the left.

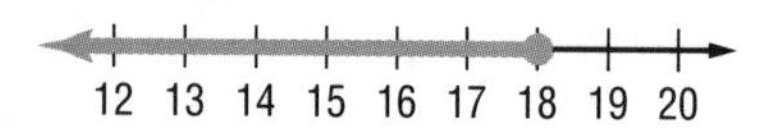

4. Check your graph. Choose a number from the shaded part of the line.

$14 \leq 18$ ✔

YOUR TURN!

Graph the inequality.
$x \geq 64$

1. Draw a number line that includes ______.
2. The point at ______ will be a ________ circle. Draw the point.
3. The inequality includes numbers ____________ 64, so shade the line to the ________.

61 62 63 64 65 66 67 68 69

4. Check your graph. Choose a number from the shaded part of the line.

_____ ≥ 64

Example 2

Write the inequality shown on the number line.

–19–18–17–16–15–14–13–12–11

1. The number at the point is 12.
2. The shaded part of the line has numbers less than 12.
3. The circle is open so the inequality symbol is <.
4. Write the inequality. $x < 12$

YOUR TURN!

Write the inequality shown on the number line.

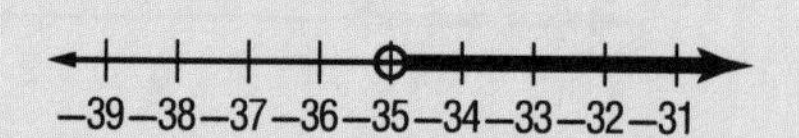

1. The number at the point is _____.
2. The shaded part of the line has numbers ________________.
3. The circle is ________ so the inequality symbol is _____.
4. Write the inequality. x ◯ _____

Guided Practice

Graph each inequality.

1. $k < -12$

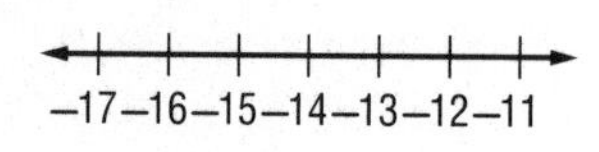

2. $x > -3$

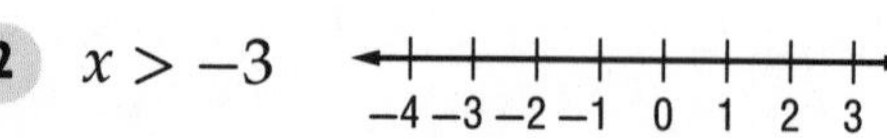

Step by Step Practice

3 Write the inequality shown on the number line.

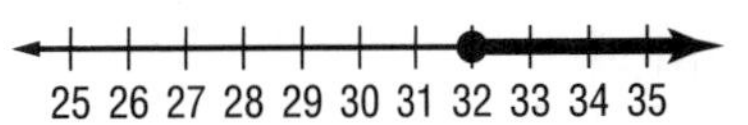

Step 1 The point is at ____. The circle is __________.

Step 2 The shaded part of the line has numbers __________ than 32.
The circle is closed so the inequality symbol is ____.

Step 3 Write the inequality.

n 

Write each inequality.

4

−5 −4 −3 −2 −1 0 1 2 3 4 5

x 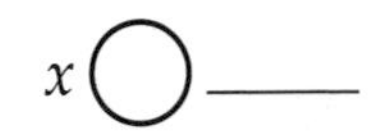

5

−5 −4 −3 −2 −1 0 1 2 3 4 5

b 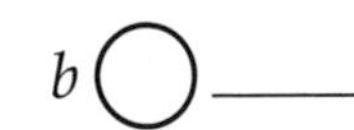

Step by Step Problem-Solving Practice

Solve.

6 Jarrod mows grass to earn money. He needs to earn at least $375 to buy a set of golf clubs. Write an inequality that names the amount he wants.

Let g = ______________. $375 is the ______________

amount of money Jarrod wants. For *at least,* use the symbol ____.

350 375 400

Check off each step.

_____ **Understand: I underlined key words.**

_____ **Plan: To solve the problem, I will** ______________________.

_____ **Solve: The answer is** ______________________.

_____ **Check: I checked my answer by** ______________________.

GO ON

Skills, Concepts, and Problem Solving

Graph each inequality.

7. $b > -25$

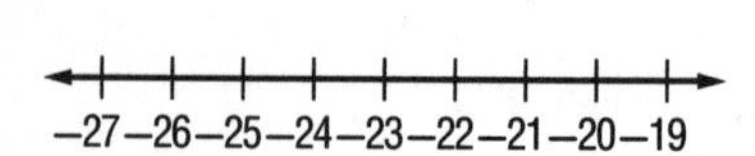

8. $h > 48$

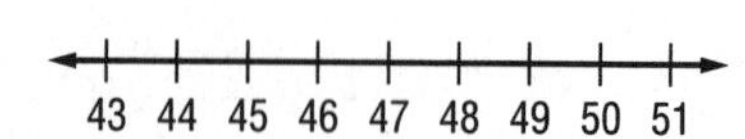

9. $x \leq -12$

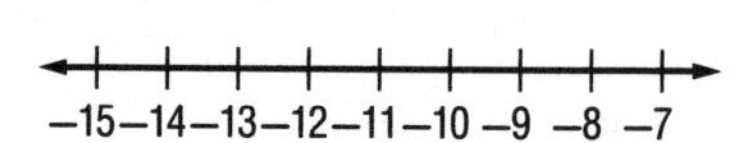

10. $j \geq 37$

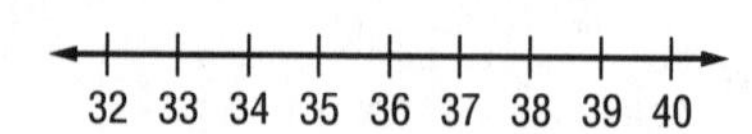

Write each inequality.

11.

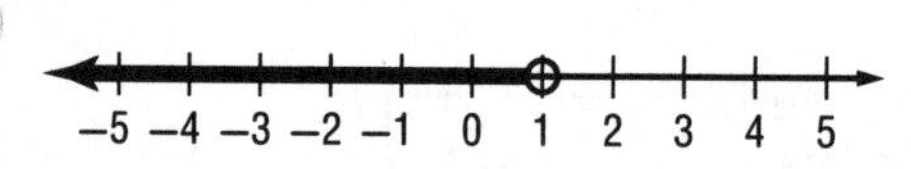

____

12.

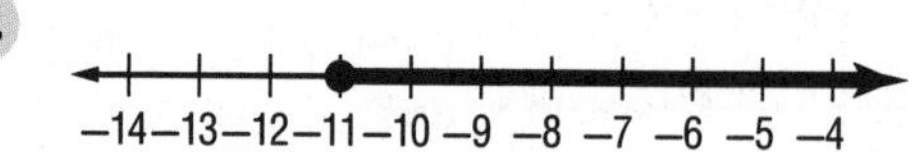

____

Solve.

13. **SAFETY** The local community lodge has a maximum occupancy of 480 people. Write an inequality to show the maximum occupancy.

____

14. **GEOMETRY** An obtuse angle has a measure of greater than 90°. Write an inequality to show the measurement of an obtuse angle.

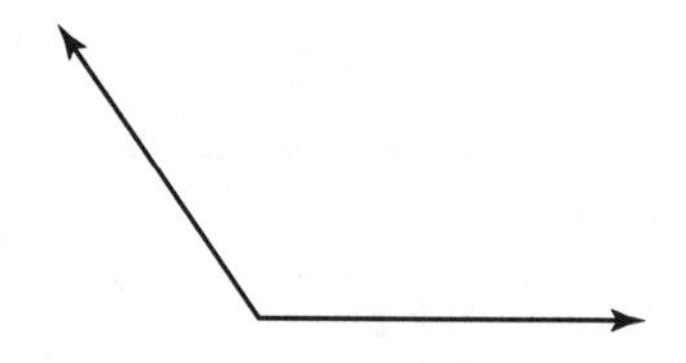

____

Vocabulary Check **Write the vocabulary word that completes each sentence.**

15. A(n) ____________ is a number sentence that compares two unequal expressions and uses $>$, $<$, $\geq$, $\leq$, or $\neq$.

16. A ____________ makes an inequality or equation true.

17. **Reflect** Describe the difference between $x \leq 6$ and $x < 6$.

____

____

STOP

Chapter 4 Progress Check 4 (Lessons 1-7 and 1-8)

Solve each equation.

1 $r + 4 = 13$

$r =$ ______

2 $31 = -8 + v$

$v =$ ______

3 $x - 1 = -4$

$x =$ ______

4 $\frac{x}{6} = -4$

$x =$ ______

5 $7x = 35$

$x =$ ______

6 $\frac{x}{-2} = 4$

$x =$ ______

7 $11 + a = 16$

$a =$ ______

8 $8c = -48$

$c =$ ______

9 $d - 14 = 14$

$d =$ ______

Graph each inequality.

10 $g < -13$

–20 –19 –18 –17 –16 –15 –14 –13 –12 –11 –10

11 $z > 55$

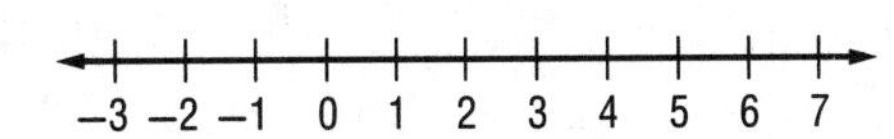

12 $y \geq -10$

–10 –9 –8 –7 –6 –5 –4 –3 –2 –1 0

13 $w \leq 7$

–3 –2 –1 0 1 2 3 4 5 6 7

Write each inequality.

14

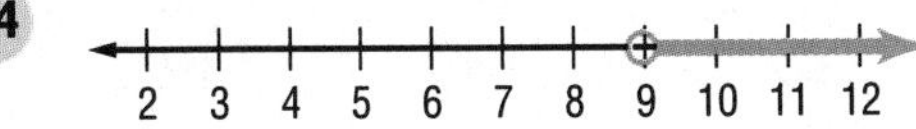

______

15

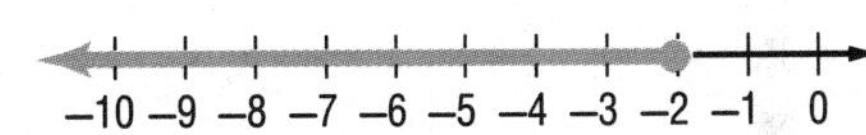

______

Solve.

16 **SCHOOL** The school auditorium has a maximum occupancy of 125 people. Write an inequality to show the maximum occupancy.

______

17 **COLORING** Clara and Jenna are buying crayons to share with two of their friends. They buy a package that has 64 crayons. How many crayons will they give their friends if they each keep the same number of crayons for themselves?

______

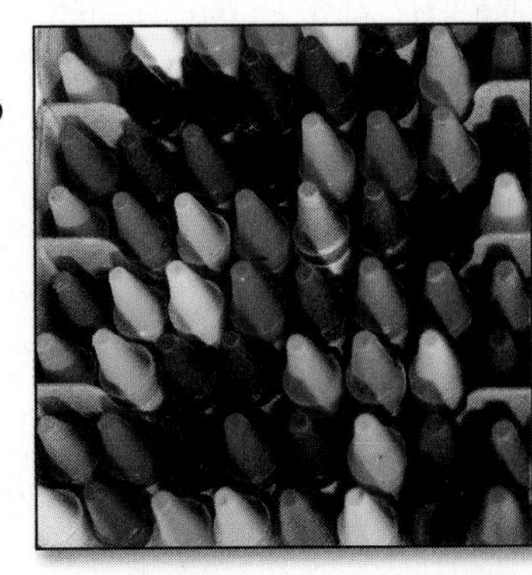

Chapter Test

Name all sets in which each number belongs.

1 -36 ________

2 0 ________

Name the opposite of each number.

3 -7 ______

4 1 ______

5 65 ______

6 -115 ______

Order each list from least to greatest.

7 34, −41, −14, 43, 143

____

8 1,809; −1,980; 1,998; −899; 1,008

____

Find each sum or difference.

9 $-90 + 18 =$ ______

10 $22 + (-11) =$ ______

11 $-6 - (-24) =$ ______

12 $64 - (-88) =$ ______

13 $31 + 163 =$ ______

14 $96 + (-48) =$ ______

15 $1 + (-2) + 20 =$ ______

16 $-4 - (-6) + 50 =$ ______

17 $-10 - 33 + (-59) =$ ______

Find each product or quotient.

18 $-7 \cdot 12 =$ ______

19 $4 \cdot (-15) =$ ______

20 $-28 \div (-7) =$ ______

21 $11 \cdot (-45) =$ ______

22 $-126 \div 3 =$ ______

23 $100 \cdot (-80) =$ ______

Solve each equation.

24 $x + 8 = -12$

$x =$ ______

25 $56 = -8w$

$w =$ ______

26 $62 = h - 10$

$h =$ ______

27 $b - 14 = 90$

$b =$ ______

28 $34 = \frac{n}{2}$

$n =$ ______

29 $-5 = \frac{y}{3}$

$y =$ ______

Graph each inequality.

30 $v > -22$

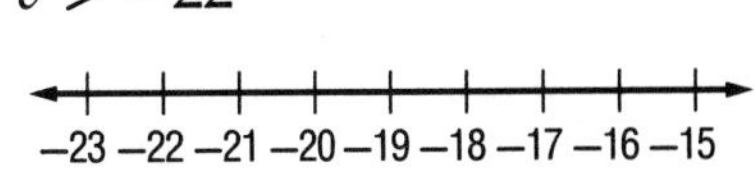

31 $k < -18$

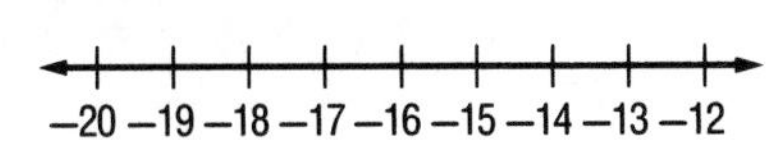

32 $x \leq 20$

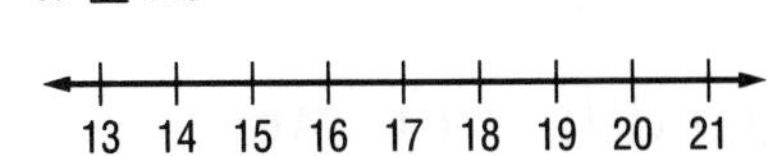

33 $d \leq 17$

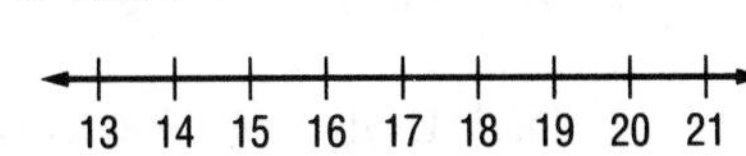

Write each inequality.

34 –10 –9 –8 –7 –6 –5 –4 –3 –2 –1 0

____

35

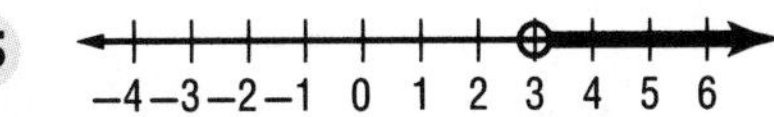

____

Solve.

36 **FINANCE** The stock market plunged 120 points over the last 4 days. What is the average daily change?

____

37 **THEATER** The fall musical's attendance increased each night of the students' performance. Which night had the most people in attendance? Which night had the least?

____

Wednesday	Thursday	Friday	Saturday	Sunday
135	122	205	233	250

Correct the mistake.

38 Adrian said that if a multiplication problem has a negative factor, then the product must be negative. Give an example of a multiplication sentence that has a negative factor and does not have a negative product. Give an example of a multiplication sentence that has only one negative factor and does not have a negative product.

____

____

STOP

Chapter 2 Decimals

Batting Averages

Batting averages are rounded to thousandths. You can use place value to compare and order the batting averages of different baseball players.

STEP 1 Chapter Pretest Are you ready for Chapter 2? Take the Chapter 2 Pretest to find out.

STEP 2 Preview Get ready for Chapter 2. Review these skills and compare them with what you will learn in this chapter.

What You Know

You know the place value of decimals.

3.15

The digit 1 is in the tenths place. The value of the digit is 0.1.

TRY IT!

Name the place of the underlined digit. Then write the value of the digit.

1. 43.26 ____________
2. 3.682 ____________
3. 82.601 ____________
4. 0.657 ____________

What You Will Learn

Lesson 2-3

To round a decimal, first underline the digit to be rounded.

Then look at the digit to the right of the place being rounded.

Example:
Round 1.78 to the nearest whole number.

Underline the digit to be rounded.

1.78

Since the digit is 7, add one to the underlined digit.

On the number line, 1.7 is closer to 2.0 than to 1.0. To the nearest whole number, 1.78 rounds to 2.

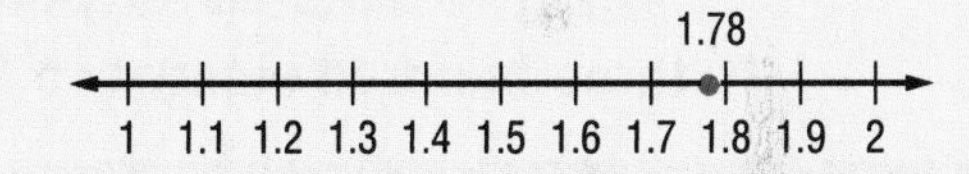

What You Know

You know how to add multi-digit numbers.

Example:

$$\begin{array}{r} {}^{1}\\ 352 \\ +\ 152 \\ \hline 504 \end{array}$$

TRY IT!

Add.

5. $\begin{array}{r} 649 \\ +\ 213 \\ \hline \end{array}$

6. $\begin{array}{r} 291 \\ +\ 439 \\ \hline \end{array}$

What You Will Learn

Lesson 2-4

To add or subtract decimals, line up the decimal points.

Then, add or subtract digits in the same place-value position.

45.2	Line up the decimal points.
+ 1.5	Add.
46.7	

Lesson 2-1

Decimals and Place Value

KEY Concept

The base 10 number system is built on place values of 10. To read and write decimals, use a place-value chart.

1000	100	10	1	0.1	0.01	0.001	0.0001
thousands	hundreds	tens	ones	tenths	hundredths	thousandths	ten-thousandths
	5	6	1 .	3	8	2	
	500	60	1	0.3	0.08	0.002	

3 tenths or 0.3

8 hundredths or 0.08

2 thousandths or 0.002

The decimal point separates the whole number part from the fraction part. It is read "**and**."

The number in the chart is read:

561.382

five hundred sixty-one and three hundred eighty-two thousandths

VOCABULARY

decimal
a number that can represent whole numbers and fractions

decimal point
point separating the ones and tenths in a number

place value
the value given to a digit by its position in a number

When a decimal is read correctly, the fraction is named.

For example, 0.23 is twenty-three hundredths or $\frac{23}{100}$.

Example 1

Name the place value of the underlined digit in 4.09<u>2</u>.

1. The underlined digit is 2. It is three places to the right of the decimal point.

2. The third digit to the right of the decimal point is the thousandths place.

YOUR TURN!

Name the place value of the underlined digit in 180.<u>6</u>09.

1. The underlined digit is ________.

   It is ________ place to the ________ of the decimal point.

2. The ________ digit to the ________ of the decimal point is the ________ place.

Example 2

Write 8,056.011 in the place-value chart.

1. Count the number of digits to the left of the decimal point. There are 4 digits.
2. Begin in the thousands column and fill in the 4-digit whole number.
3. Count the number of digits to the right of the decimal point. There are 3 digits.
4. Begin in the tenths column and fill in the 3-digit decimal number.

1000	100	10	1	0.1	0.01	0.001	0.0001
thousands	hundreds	tens	ones	tenths	hundredths	thousandths	ten-thousandths
8	0	5	6.	0	1	1	

YOUR TURN!

Write 989.406 in the place-value chart.

1. Count the number of digits to the ____________ of the decimal point. There are ____________ digits.
2. Begin in the ____________ column and fill in the ____________ whole number.
3. Count the number of digits to the ____________ of the decimal point. There are ____________ digits.
4. Begin in the tenths column and fill in the ____________ decimal number.

1000	100	10	1	0.1	0.01	0.001	0.0001
thousands	hundreds	tens	ones	tenths	hundredths	thousandths	ten-thousandths

Example 3

Write the phrase as a decimal.

four hundred and eight-five thousandths

1. Read the number before "and." **400**
2. The last word "thousandths" indicates three places are after the decimal point.
3. Write 400 followed by a decimal point and three blanks for the decimal part.

   400. _ _ _
4. Read the number after "and." **85**
5. Begin at the right to fill in the blanks. Fill the last blank with 0.

   400. **0 8 5**

YOUR TURN!

Write the phrase as a decimal.

sixteen and fifteen hundredths

1. Read the number before "and." ______
2. The last word "____________" indicates ______ places are after the decimal point.
3. Write ______ followed by a decimal point and ____________ for the decimal part.

   ______.______
4. Read the number after "and." ______
5. Begin at the right to fill in the blanks.

   ______.______.

GO ON

Guided Practice

Name the place value of each underlined digit.

1. 47.5<u>1</u>3

   The underlined digit is in the ____________________ place.

   The underlined digit is ____________________.

2. 802.601<u>5</u>

   The underlined digit is in the ____________________ place.

   The underlined digit is ____________________.

Write each number in the place-value chart.

3. 37.0915

1000	100	10	1	0.1	0.01	0.001	0.0001
thousands	hundreds	tens	ones	tenths	hundredths	thousandths	ten-thousandths

4. 0.45

1000	100	10	1	0.1	0.01	0.001	0.0001
thousands	hundreds	tens	ones	tenths	hundredths	thousandths	ten-thousandths

Step by Step Practice

5. Write four and nine tenths as a decimal.

   Step 1 Write the number for the words ________ "and." ________

   Step 2 Write the number for the words ________ "and." ________

   Step 3 Combine the two numbers, separate by a period, to write the decimal. ______

Write the phrase as a decimal.

6. fifty-one and thirteen thousandths

   Write the number for the words __________ "and." __________

   Write the number for the words __________ "and." __________

   Combine the two parts to write the decimal. __________

Step by Step Problem-Solving Practice

Solve.

7 **STOCK MARKET** The NASDAQ Stock Market closed at 2,408.04. Write the closing number in word form.

1000	100	10	1	0.1	0.01	0.001	0.0001
thousands	hundreds	tens	ones	tenths	hundredths	thousandths	ten-thousandths

____________________ and ____________________

Check off each step.

______ Understand: I underlined key words.

______ Plan: To solve the problem, I will ____________________.

______ Solve: The answer is ____________________.

______ Check: I checked my answer by ____________________.

Skills, Concepts, and Problem Solving

Name the place value of each underlined digit.

8 1,<u>6</u>22.97

____

9 0.00<u>7</u>

____

Write each number in the place-value chart.

10 1.054

1000	100	10	1	0.1	0.01	0.001	0.0001
thousands	hundreds	tens	ones	tenths	hundredths	thousandths	ten-thousandths

11 380.7

1000	100	10	1	0.1	0.01	0.001	0.0001
thousands	hundreds	tens	ones	tenths	hundredths	thousandths	ten-thousandths

Write each phrase as a decimal.

12 forty-five and thirty-two hundredths

____

13 eight and sixteen ten-thousandths

____

14 one hundred two and six thousandths

____

15 sixty-eight and four tenths

____

Solve.

16 **TRACK AND FIELD** Kimberly participated in the long jump at a track meet. She jumped 16.95 feet. Write the distance she jumped in word form.

____

17 **WEIGHT** Martin stepped on the scale at the team weigh-in and his weight, in pounds, is shown on the scale. Write his weight in word form.

SCALE
126.4

____

18 **RAINFALL** Over the weekend, one and thirty-five hundredths inches of rain fell. Write the amount of rainfall in standard form.

____

Vocabulary Check **Write the vocabulary word that completes each sentence.**

19 A(n) ____________________ separates the ones and tenths in a number.

20 A number that can represent whole numbers and fractions is called

a(n) ____________________.

21 The value given to a digit by its position in a number is its

____________________.

22 **Reflect** Explain how you know where to place the decimal point when writing a number in standard form that is given in words.

____

____

STOP

Lesson 2-2 Compare and Order Decimals

KEY Concept

To compare **decimals**, you can use a number line.

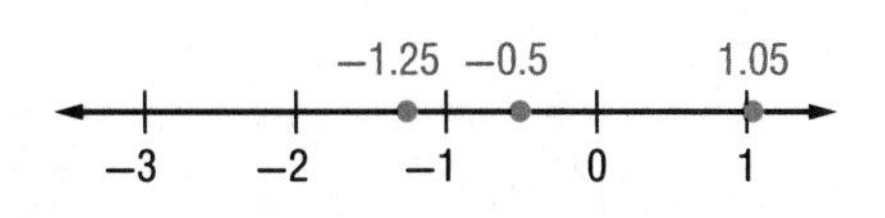

The greater number is to the right.

Use >, <, or = to compare numbers.

–1.25 < –0.5 < 1.05

–1.25 is less than –0.5

–0.5 is less than 1.05

When you use **place value** to compare and order decimals, compare digit by digit beginning on the left.

2.561

2.526

2.561 > 2.526

The digits in the ones places are equal.
The digits in the tenths places are equal.
Compare the digits in the hundredths places.
6 > 2

VOCABULARY

inequality
a number sentence that compares two unequal expressions and uses <, >, ≤, ≥, or ≠

decimal
a number that can represent whole numbers and fractions

place value
a value given to a digit by its position in a number

When comparing negative decimals, the greatest decimal is the decimal closest to zero.

Example 1

Use >, <, or = to compare 0.82 and 0.52.

1. Graph the numbers on a number line.

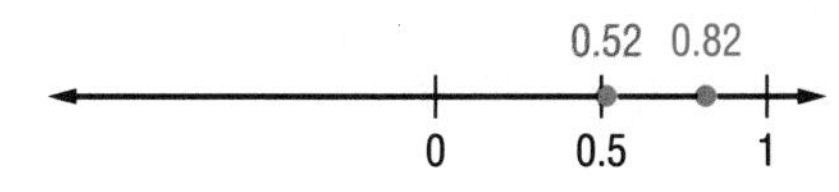

2. Because 0.82 is to the right of 0.52, it is the greater number.

3. Write an inequality statement.

0.82 > 0.52

YOUR TURN!

Use >, <, or = to compare –0.75 and –0.79.

1. Graph the numbers on a number line.

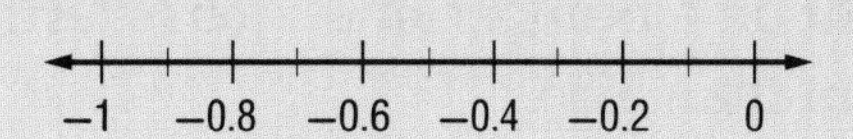

2. Because ________ is to the ________ of ________, it is the ________ number.

3. Write an inequality statement.

–0.75 ◯ –0.79

GO ON

Example 2

Use $>$, $<$, or $=$ to compare 4.505 and 4.507.

1. Compare the ones digits.

   $4 = 4$

2. Compare the digits in the tenths places.

   $5 = 5$

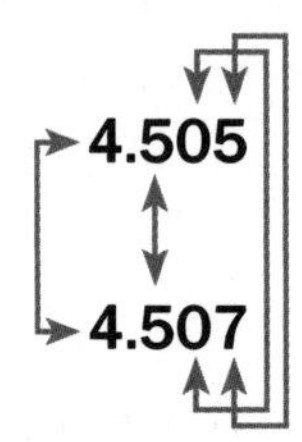

3. Compare the digits in the hundredths places.

   $0 = 0$

4. Compare the digits in the thousandths places.

   $5 < 7$

5. Write an inequality statement.

   $4.505 < 4.507$

YOUR TURN!

Use $>$, $<$, or $=$ to compare 81.327 and 81.237.

1. Compare the _______ digits.

   ____

2. Compare the _______ digits.

   ____

3. Compare the digits in the _______ places.

   ____

4. Write an inequality statement.

   81.327 ◯ 81.237

Example 3

Order the decimals 3.56, 3.65, −3.60, 5.05, and −6.55 from least to greatest.

1. For the negative decimals, −6.55 is farther to the left than −3.60 on a number line.

   $-6.55 < -3.60$

2. For the positive decimals, 5.05 is farthest to the right on a number line. 3.65 is farther to the right than 3.56.

   $3.56 < 3.65 < 5.05$

3. Write the decimals from least to greatest.

   −6.55, −3.60, 3.56, 3.65, 5.05

YOUR TURN!

Order the numbers −0.23, 0.701, −2.37, 0.007, and 3.133 from least to greatest.

1. For the negative numbers, _______ is farther to the right than _______ on a number line.

   _______ $<$ _______

2. For the positive decimals, _______ is farthest to the right on a number line. _______ is farther to the right than _______.

   _______ $<$ _______ $<$ _______

3. Write the decimals from least to greatest.

   ____

Guided Practice

Use >, <, or = to compare each pair of numbers.

1. 2.54 ◯ 2.45

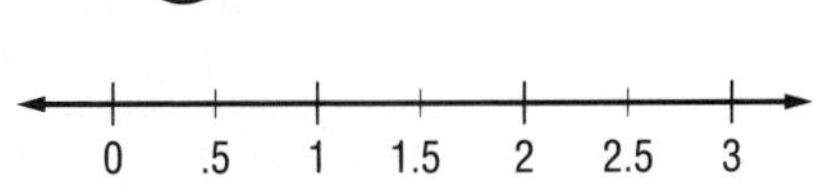

2. 0.185 ◯ 0.189

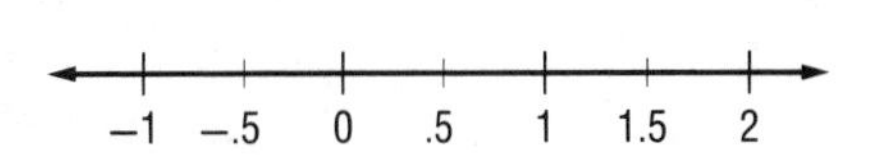

3. 463.0666 ◯ 463.066

In which place values are the digits the same?

____

After comparing the digits, which place value helped decide which inequality symbol to use? ______________________

4. 11.12 ◯ 11.112

After comparing the digits, which place value helped decide which inequality symbol to use? ______________________

Step by Step Practice

5. Order the numbers −2.65, 1.56, −0.256, −2.15, and 0.615 from least to greatest.

Step 1 For the negative numbers, ________ is to the left of both ________ and ________ on a number line. ________ is left of ________.

________ < ________ < ________

Step 2 For the positive numbers, ________ is to the right of ________ on a number line.

________ < ________

Step 3 Write the numbers from least to greatest.

________, ________, ________, ________, ________

GO ON

Order each set of numbers from least to greatest.

6 2.86, 2.68, 2.068, 2.086, 0.286

____

7 −15.0765, −15.6570, −15.5067, −15.0675, −15.6057

____

Step by Step Problem-Solving Practice

Solve.

8 **DRIVING** There are three different driving routes for Will to get to school. The distance for Route A is 1.4 miles. The distance for Route B is 1.26 miles. The distance for Route C is 1.34 miles. Which route distance is the shortest?

Compare the digits in the ones places. 1.4, 1.34, 1.26

____

Compare the digits in the tenths places. 1.4, 1.34, 1.26

____

The distances in order from least to greatest are

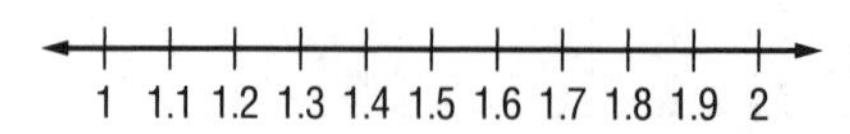

_______________.

Check off each step.

_____ **Understand: I underlined key words.**

_____ **Plan: To solve the problem, I will** _______________.

_____ **Solve: The answer is** _______________.

_____ **Check: I checked my answer by** _______________.

Skills, Concepts, and Problem Solving

Use >, <, or = to compare each pair of numbers.

9 1.22 ◯ 1.02

10 6.28 ◯ 6.279

11 64.23 ◯ 64.32

12 187.2 ◯ 187.200

13 6.20 ◯ 6.04

14 0.99 ◯ 1.1

15 0.554545 ◯ 0.554554

16 7.0058 ◯ 7.058

17 0.12 ◯ 0.1200

Order each set of numbers from least to greatest.

18 71.4, −47.1, −0.0771, 4.17

____

19 −1.0095, −15.0909, −15.0099, 5.0919, −15.9017

____

20 −4.0041, −0.441, −1.4, 14.0111, 44.001

____

Solve.

21 **MARATHON** Alana and her sisters, Danielle and Crystal, ran the city marathon. Their times are shown in the table at the right. Order the sisters' times from fastest to slowest.

Runner	Alana	Danielle	Crystal
Time	4.62 hr	4.97 hr	4.87 hr

____

22 **CHEMISTRY** A sample had a mass of 2413.11 grams at the start of a day and had a mass of 2408.04 grams at the end of the day. Did the sample weigh more in the morning or in the afternoon?

____

23 **BATTING AVERAGE** After the first 15 games, Jack's batting average was 0.378. After 30 games, his batting average changed to 0.387. Did his batting average improve?

____

Vocabulary Check **Write the vocabulary word that completes each sentence.**

24 A(n) ____________________ uses $<, >, \leq, \geq$, or $\neq$ to compare two expressions.

25 A value given to a digit by its position in a number is the

____________________.

26 **Reflect** Give a number that is between 1.23 and 1.24. Explain how you know that it is between the two numbers.

____

____

STOP

Chapter 2

Progress Check 1 (Lessons 2-1 and 2-2)

Write each number in a place-value chart.

1 460.087

1000	100	10	1	0.1	0.01	0.001	0.0001
thousands	hundreds	tens	ones	tenths	hundredths	thousandths	ten-thousandths

2 1,295.3

1000	100	10	1	0.1	0.01	0.001	0.0001
thousands	hundreds	tens	ones	tenths	hundredths	thousandths	ten-thousandths

Write the value of each underlined digit.

3 $3{,}0\underline{2}0.005$

____

4 $5.\underline{2}64$

____

5 $500.10\underline{4}7$

____

6 $719.0\underline{4}5$

____

Use >, <, or = to compare each pair of numbers.

7 7.34 ◯ 6.9

8 5.05 ◯ 5

9 72.45 ◯ −74

10 39.04 ◯ 39.4

11 0.887 ◯ −0.098

12 3.24 ◯ 3.240

Order the set of numbers from least to greatest.

13 63.1, −36.9, −9.36, 33.6, 0.39

____

Solve.

14 **ADDRESS** Chen lives at 507 Washington Street. Write his address number in word form.

____

15 **MOVIES** Sarita has three movies she wants to watch. She plans to watch them in order of longest to shortest movie time. Write the order she will watch the movies.

____

____

Movies	Time
The Sound of Music	174 min
The Princess Bride	98 min
101 Dalmatians	103 min

Lesson 2-3

Round Decimals

KEY Concept

To **round** a decimal, you can use a number line.

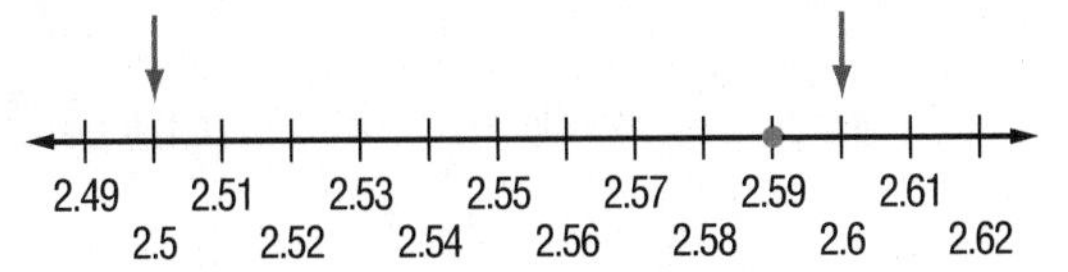

To the nearest tenth, 2.59 rounds to 2.6.

To round a decimal without a number line, look at the digit to the right of the place being rounded.

To the nearest hundredth,

1.567 rounds to 1.57

place you are rounding to — 7 > 5, so add 1 to the 6. Drop all digits to the right of 6.

To the nearest hundredth,

2.493 rounds to 2.49

place you are rounding to — 3 < 5, so drop all digits to the right of 9.

VOCABULARY

decimal
a number that can represent whole numbers and fractions

place value
the value given to a digit by its position in a number

round
to find the nearest number based on the given place value

Round decimals when you are estimating an answer or when exact numbers are not needed.

Example 1

Round 4.<u>8</u>3 to the place value of the underlined digit.

1. Graph 4.83 on a number line.

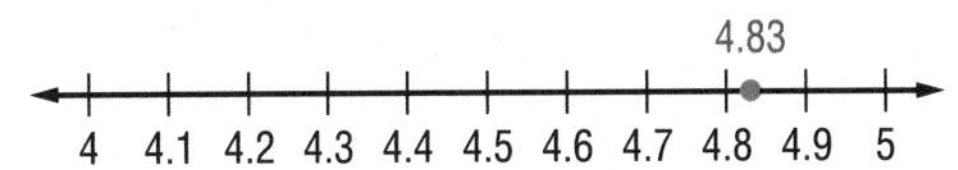

2. 4.83 is between 4.8 and 4.9, but closer to 4.8.

3. To the nearest tenth, 4.83 rounds to 4.8.

YOUR TURN!

Round 2.<u>1</u>6 to the place value of the underlined digit.

1. Graph 2.16 on a number line.

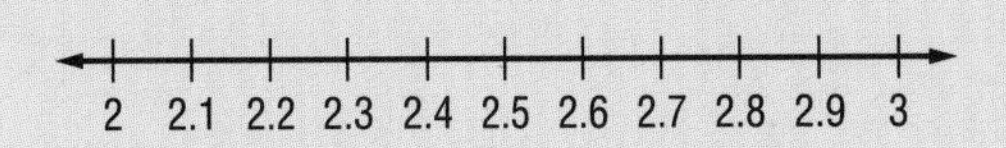

2. 2.16 is between _____ and _____, but closer to _____.

3. To the nearest tenth, 2.16 rounds to _____.

GO ON

Example 2

Round 14.095 to the nearest hundredth.

1. Underline the digit in the hundredths place.

   $14.0\underline{9}5$

2. Compare the digit to the right of the 9 to 5.

   $5 = 5$

3. Because $5 = 5$, add 1 to the 9 and drop digits to the right of the 9.

   $14.1\underline{0}$

   Adding 1 to 9 makes 10, which means there will be a 0 in the hundredth place and a 1 in the tenths place.

4. 14.095 rounds to 14.10.

YOUR TURN!

Round 43.127 to the nearest tenth.

1. Underline the digit in the ________ place.

   ____

2. Compare the digit to the right of the ________ to 5.

   ________ ◯ ________

3. Because ________, drop digits to the right of the ________.

4. 43.127 rounds to ________.

Guided Practice

Round each decimal to the place value of the underlined digit.

1. $7.\underline{2}9$ ______

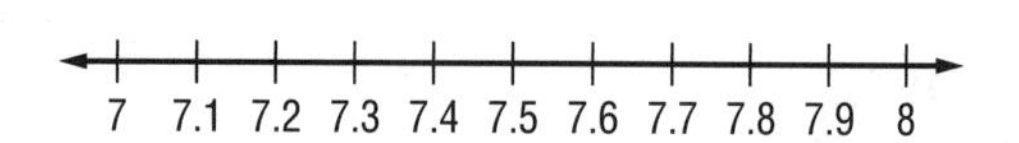

7.29 is closer to ______ than to ______,

so round ____________.

2. $9.\underline{4}15$ ______

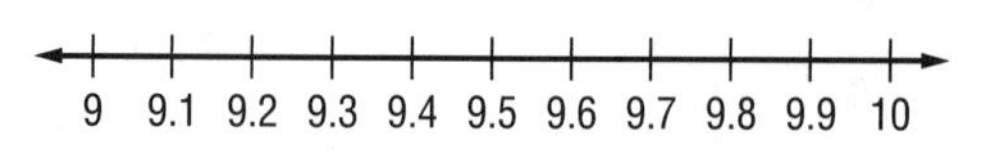

9.415 is closer to ______ than to ______,

so round ____________.

3. $0.\underline{5}55$ ______

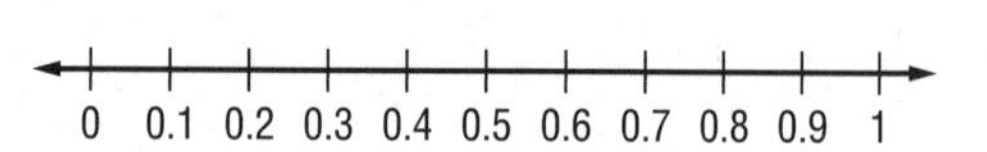

0.555 is closer to ______ than to ______,

so round ____________.

4. $59.\underline{6}3$ ______

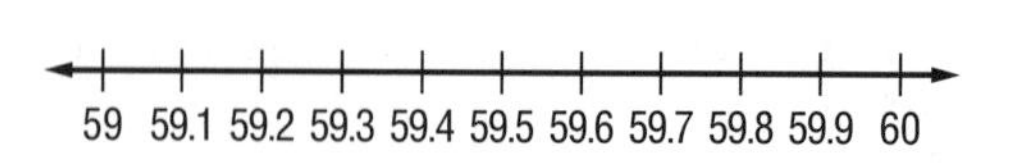

59.63 is closer to ______ than to ______,

so round ____________.

Step by Step Practice

5 Round 48.41 to the nearest tenth.

Step 1 Underline the digit in the tenths place. 48.41

Step 2 Draw an arrow to the digit at which you need to look. Compare to 5. 1 ◯ 5

Step 3 The 4 ______________. Drop all digits after the ______________.

48.41 rounded to the nearest ________ is ________.

Round each decimal to the nearest hundredth.

6 1.806 ______________

What digit is in the hundredths place? ______________

Look to the ______________ place.

7 425.599 ______________

What digit is in the hundredths place? ______________

Look to the ______________ place.

Round each decimal to the nearest one.

8 54.32 ______________

9 194.88 ______________

Round each decimal to the nearest tenth.

10 0.261 ______________

11 278.045 ______________

Step by Step Problem-Solving Practice

Solve.

12 **EDUCATION** Mrs. Roswell calculated the class average on the last exam to be 82.6815. Round the class average to the nearest tenth.

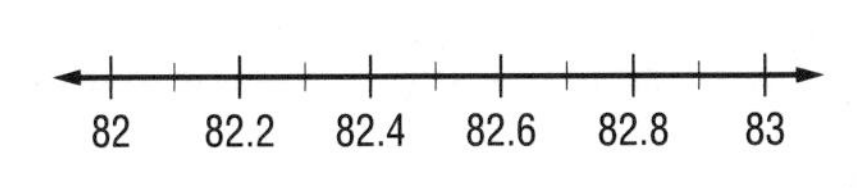

Check off each step.

______ **Understand: I underlined key words.**

______ **Plan: To solve the problem, I will** ______________________________.

______ **Solve: The answer is** ______________________________.

______ **Check: I checked my answer by** ______________________________.

GO ON

Skills, Concepts, and Problem Solving

Round each number to the place value of the underlined digit.

13 $578.9\underline{2}6$ ________

14 $0.\underline{7}29$ ________

15 $1\underline{8}.7922$ ________

16 $\underline{8}72.002$ ________

17 $0.02\underline{3}4$ ________

18 $\underline{5}3.828$ ________

Round 207.3296 to each place value shown.

19 tenth ________

20 hundred ________

21 ten ________

22 one ________

23 hundredth ________

24 thousandth ________

Solve.

25 **MONEY** Ryan wanted to buy a birthday present for his mother. He has $59.38. Round his money to the nearest dollar.

____

26 **PI** The mathematical constant *pi* (or π), 3.14159265358979323846…, represents the ratio of a circle's circumference to its diameter. Round *pi* to the nearest thousandth.

____

27 **AUTO RACING** In NASCAR racing, Kyle Busch was the mile leader with 672.15 miles completed. Round the number of miles completed to the nearest tenth.

____

Vocabulary Check **Write the vocabulary word that completes each sentence.**

28 ________________ is the value given to a digit by its position in a number.

29 To ________________ is to find the nearest number based on the given place value.

30 **Reflect** Summarize how to round a number.

____

____

____

STOP

Lesson 2-4

Add and Subtract Decimals

KEY Concept

These models show decimals to the hundredths.

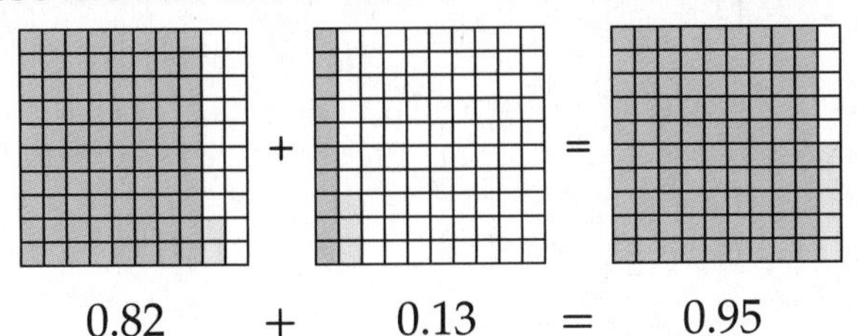

0.82 + 0.13 = 0.95

Combine the tenths. Combine the hundredths.

$0.82 + 0.13 = 0.95$

To add or subtract decimals without a model, line up the **decimal points**. Place zeros so that each number has the same number of decimal places.

13.098 + 9.3 + 22.19

```
   1
  13.098
   9.300
+ 22.190
  44.588
```

Bring the decimal straight down to place in the sum.

1.4 − 0.68

```
 0 13 10
  1.40
- 0.68
  0.72
```

Bring the decimal straight down to place in the difference.

VOCABULARY

difference
the answer in a subtraction problem

decimal
a number that can represent whole numbers and fractions

decimal point
point separating the ones and tenths in a number

sum
the answer in a addition problem

You can use decimal models to subtract decimals. Model the first number, then take away the second.

Example 1

Find the sum of 0.73 and 0.5 using decimal models.

1. Model each decimal.

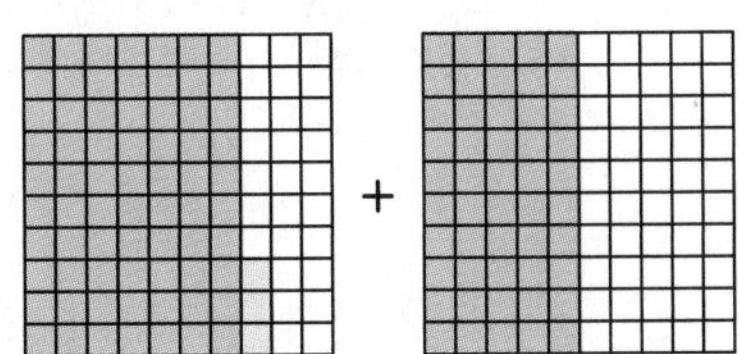

2. Count the total number of shaded blocks.

0.73 + 0.5 = 1.23

YOUR TURN!

Find the sum of 0.38 and 0.11 using decimal models.

1. Model each decimal.

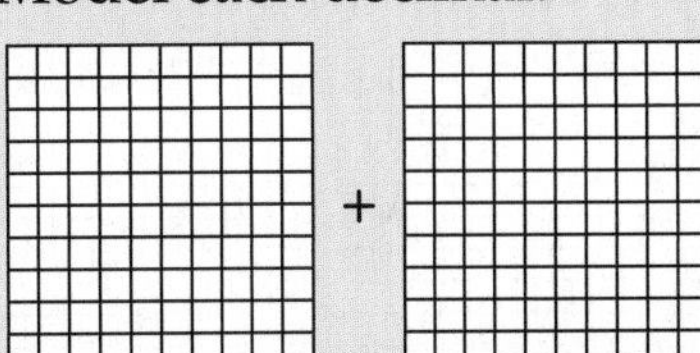

2. Count the total number of shaded blocks.

0.38 + 0.11 = ________

Example 2

Find 1.23 – 0.9 using decimal models.

1. Model the first decimal, 1.23.
2. Mark out 0.9 or 9 tenths.
3. Count the blocks that are left.

1.23 – 0.9 = 0.33

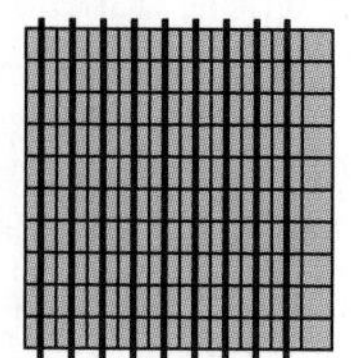 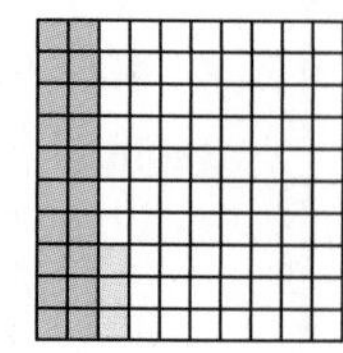

YOUR TURN!

Find 0.63 – 0.08 using decimal models.

1. Model the first decimal, 0.63.
2. Mark out 0.08 or ________.
3. Count the blocks that are left.

0.63 – 0.08 = ________

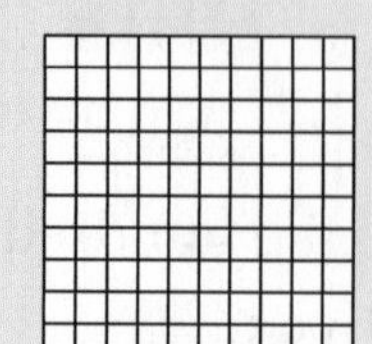

Example 3

Find 83.019 – 45.92.

1. Write the numbers vertically.
   Line up the decimal points.
   Write zeros as placeholders.
2. Subtract.

$$\begin{array}{r} 83.019 \\ -\ 45.920 \\ \hline 37.099 \end{array}$$

3. Place the decimal point in the difference.

83.019 – 45.92 = 37.099

YOUR TURN!

Find 14.078 + 86.902.

1. Write the numbers vertically.
   Line up the decimal points.
   Write ________ as placeholders.
2. Add.

   \+ ________

3. Place the decimal point in the ________.

   14.078 + 86.902 = ________

Guided Practice

Find each sum using decimal models.

1. 0.34 + 0.57 = ________

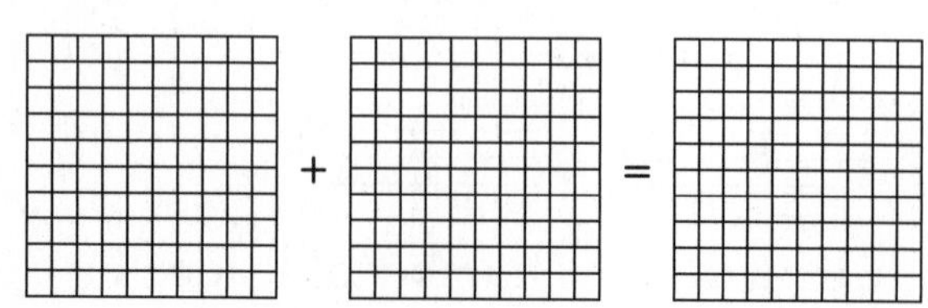

2. 0.09 + 0.37 = ________

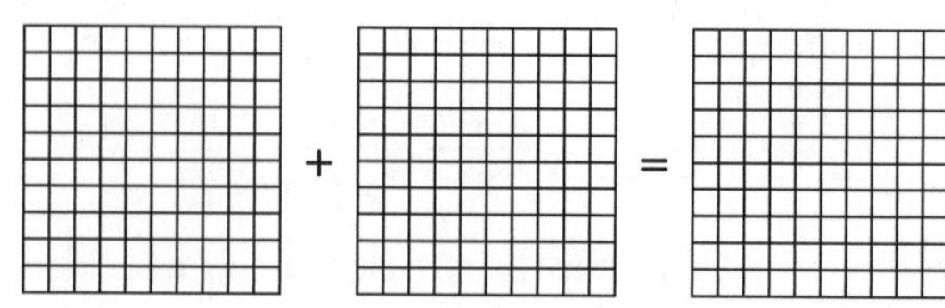

Find each difference using decimal models.

3 $1.22 - 0.51 =$ ____________

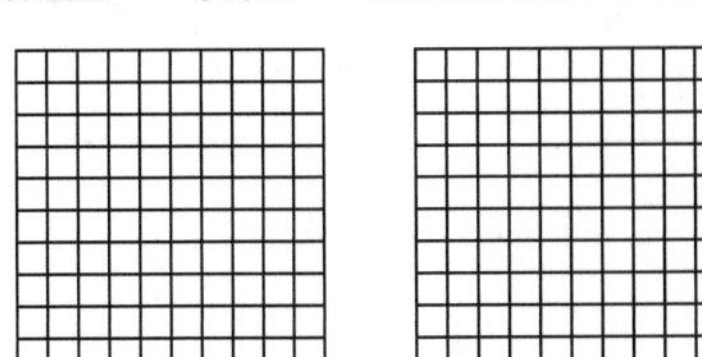

4 $0.79 - 0.16 =$ ____________

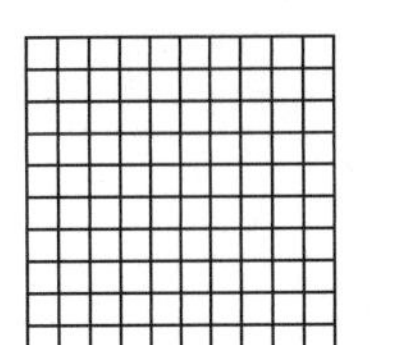

Step by Step Practice

5 Find the sum of 16.438 and 5.21.

Step 1 Write the numbers vertically. Line up the decimal points. Write zeros as placeholders when necessary.

Step 2 Add.

Step 3 Place the decimal point in the sum.

Find each sum or difference.

6 $3.2 - 1.32 =$ ____________

7 $0.48 + 3.701 =$ ____________

Step by Step Problem-Solving Practice

Solve.

8 **SWIMMING** The Dolphins Swim Team had practice three days this week. They practiced for 3.5 hours Tuesday, 2.25 hours Thursday, and 4 hours Saturday. How many hours did they practice during the week?

+ ______

Check:

$-$ 3.50

$-$ 2.25

$-$ 4.00

Check off each step.

______ Understand: I underlined key words.

______ Plan: To solve the problem, I will ____________________.

______ Solve: The answer is ____________________.

______ Check: I checked my answer by ____________________.

GO ON

Skills, Concepts, and Problem Solving

Find each sum or difference.

9. 5.86 − 3.416 = __________

10. 7.029 + 0.32 = __________

11. 6.99 − 3.7 = __________

12. 31.8 + 15.605 = __________

13. 7.24 + 5.18 = __________

14. 58.21 − 9.6 = __________

Solve.

15. **MEASUREMENT** Dontel and Jovan were measuring the length of a textbook for a class project. Dontel measured 12.25 inches. Jovan measured 12.325 inches. What was the difference between their measurements?

____

16. **SNACKS** Renee went to the health food store for a treat. She bought 1.2 pounds of pistachios, 0.85 pounds of pumpkin seeds, and 1.62 pounds of corn nuts. How many pounds of snacks did Renee buy?

____

17. **MONEY** Cora had $15 when she went out to lunch. If she spent $5.95 on lunch, how much money does she have left?

____

Vocabulary Check **Write the vocabulary word that completes each sentence.**

18. A(n) __________________ is the answer to an addition problem.

19. A(n) __________________ is the answer to a subtraction problem.

20. In a number, the ones and tenths are separated by a(n) __________________.

21. **Reflect** When adding and subtracting decimals without a model, why is it best to write the numbers vertically?

____

____

____

STOP

Chapter 2

Progress Check 2 (Lessons 2-3 and 2-4)

Round each decimal to the place value of the underlined digit.

1. $9\underline{2}7.12$ ____________

2. $14.\underline{8}33$ ____________

3. $1\underline{3}05.423$ ____________

4. $\underline{4}99.278$ ____________

5. $4.0\underline{9}9$ ____________

6. $55\underline{8}3.1$ ____________

Round 5,836.0742 to each place value shown.

7. tenth ____________

8. hundred ____________

9. ten ____________

10. one ____________

11. hundredth ____________

12. thousandth ____________

Find each sum or difference.

13. $48.02 - 9.347 =$ ____________

14. $340.22 + 109.88 =$ ____________

15. $197.004 - 7.89 =$ ____________

16. $37.05 + 15.9 =$ ____________

17. $45.072 + 6.3 =$ ____________

18. $716 - 14.76 =$ ____________

Solve.

19. **CRAFTS** Mia and Anita each needed a piece of ribbon 24 inches long to make a bow. Each girl cut a ribbon and then measured it. Mia's ribbon was 23.4 inches. Anita's ribbon was 24.2 inches. When rounded to the nearest inch, which girl's ribbon is the wrong length?

____

20. **MONEY** Stephen's wallet contained \$22 in bills. He had \$1.46 in change in his pocket. He found \$5.15 on the floor of his car. How much money does Stephen have now?

____

Lesson 2-5

Multiply Decimals

KEY Concept

When you model multiplication of decimals less than 1, the overlapping area is the product.

$$\begin{array}{r} 0.3 \\ \times\ 0.4 \\ \hline 0.12 \end{array}$$

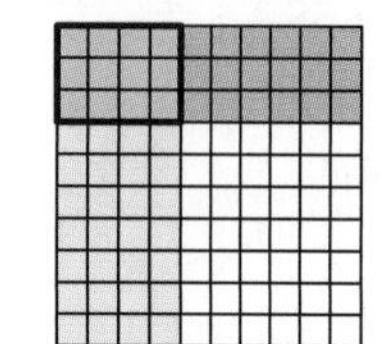

To multiply decimals without a model, place the decimal point so that the product has the same number of decimal places that the factors have combined.

$$\begin{array}{r l} 4.18 & \leftarrow \textbf{2 decimal places} \\ \times\ 2.6 & \leftarrow \textbf{1 decimal places} \\ \hline 2508 & \\ +\ 8360 & \\ \hline 10.868 & \leftarrow \textbf{3 decimal places} \end{array}$$

VOCABULARY

decimal
a number that can represent whole numbers and fractions

decimal point
point separating the ones and tenths in a number

factor
a number that is multiplied by another number in a multiplication problem

product
the answer in a multiplication problem

In the product, start at the right of the number and count left to place the decimal point.

Example 1

Find 0.5 • 0.7 using decimal models.

1. Shade 0.5 horizontally.
2. Shade 0.7 vertically.
3. Count all the shaded hundredths.

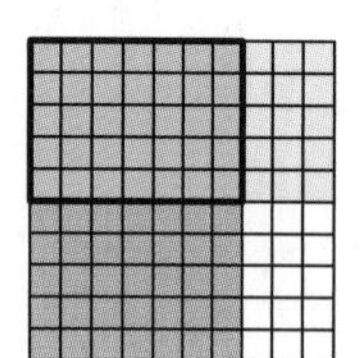

YOUR TURN!

Find 0.3 • 0.9 using decimal models.

1. Shade ______ horizontally.
2. Shade ______ vertically.
3. Count all the shaded hundredths.

0.3 • 0.9 = ______

Example 2

Find 11.9 • 0.45.

1. Write the numbers vertically.
2. Multiply.
3. Count the number of decimal places in each factor.

   $1 + 2 = 3$

4. Start at the right and move left 3 places. Place the decimal point in the product.

$$\begin{array}{r} 11.9 \\ \times\ 0.45 \\ \hline 595 \\ +\ 4760 \\ \hline 5.355 \end{array}$$

YOUR TURN!

Find 0.14 • 3.1.

1. Write the numbers vertically.
2. Multiply.
3. Count the number of decimal places in each factor.

   ______ + ______ = ______

4. Start at the right and move left ______ places. Place the decimal point in the product.

$\times$ ______

$+$ ______

Guided Practice

Find each product using decimal models.

1. 0.7 • 0.9 = __________

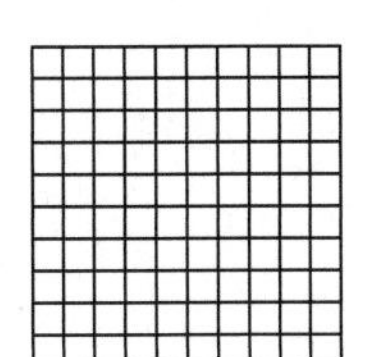

2. 0.3 • 0.5 = __________

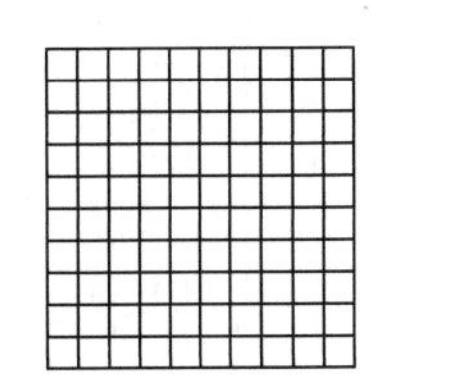

3. 1.4 • 0.6 = __________

   Shade __________ tenths vertically.

   Shade __________ tenths horizontally.

   __________ are shaded both directions.

4. 0.15 • 3 = __________

   Shade __________ tenths vertically.

   Shade __________ tenths horizontally.

   __________ are shaded both directions.

GO ON

Step by Step Practice

5 Find 6.3 • 0.24.

Step 1 Write the numbers vertically.

$$\begin{array}{r} 6.3 \\ \times\ 0.24 \\ \hline \\ +\ \quad \\ \hline \end{array}$$

Step 2 Multiply.

Step 3 Count the number of decimal places in each factor.

_____ + _____ = _____

Step 4 Start at the right and move left _____ places.

Place the decimal point in the product. _____

Find each product.

6 1.382 • 0.4 = ________

decimal places:

_____ + _____ = _____

$$\begin{array}{r} 1.382 \\ \times\ 0.4 \\ \hline \end{array}$$

7 43.7 • 0.51 = ________

decimal places:

_____ + _____ = _____

$$\begin{array}{r} 43.7 \\ \times\ 0.51 \\ \hline \\ +\ \quad \\ \hline \end{array}$$

8 7.6 • 0.56 = ________

decimal places:

_____ + _____ = _____

$$\begin{array}{r} 7.6 \\ \times\ 0.56 \\ \hline \\ +\ 3800 \\ \hline \end{array}$$

9 24.112 • 2.74 = ________

decimal places:

_____ + _____ = _____

$$\begin{array}{r} 24.112 \\ \times\ 2.74 \\ \hline \\ +\ \quad \\ \hline \end{array}$$

10 6.2 • 0.14 = ________

decimal places:

_____ + _____ = _____

$$\begin{array}{r} 6.2 \\ \times\ 0.14 \\ \hline \\ +\ \quad \\ \hline \end{array}$$

11 3.156 • 0.71 = ________

decimal places:

_____ + _____ = _____

$$\begin{array}{r} 3.156 \\ \times\ 0.71 \\ \hline \\ +\ \quad \\ \hline \end{array}$$

Step by Step Problem-Solving Practice

Solve.

12 CHARITY Cristina participated in a walk-a-thon for a local charity. She asked her sponsors to donate $0.75 for each mile she walked. If she completes 9.25 miles, how much money will each sponsor donate? Round to the nearest cent.

$$\begin{array}{r} 9.25 \\ \times\ 0.75 \\ \hline \\ +\quad\quad \\ \hline \end{array}$$

decimal places: _____ + _____ = _____

_________ ≈ _________

Check off each step.

_____ **Understand: I underlined key words.**

_____ **Plan: To solve the problem, I will** ____________________.

_____ **Solve: The answer is** ____________________.

_____ **Check: I checked my answer by** ____________________.

Estimate and check

_____ • _____ = _____

Skills, Concepts, and Problem Solving

Find each product.

13 0.5 • 0.9 = _______

14 0.3 • 0.7 = _______

15 0.8 • 1.6 = _______

16 3.2 • 0.4 = _______

17 3.4 • 2.9 = _______

18 1.7 • 2.4 = _______

19 1.09 • 0.8 = _______

20 2.34 • 0.8 = _______

21 1.56 • 4.06 = _______

GO ON

Find each product.

22 $\begin{array}{r} 18.49 \\ \times\ 2.2 \\ \hline \end{array}$

23 $\begin{array}{r} 121.78 \\ \times\ 5.1 \\ \hline \end{array}$

24 $\begin{array}{r} 3{,}005 \\ \times\ 0.43 \\ \hline \end{array}$

Solve.

25 **TRAVEL** Andre was taking a trip. He drove 3.25 hours at an average speed of 65 mph. How many miles did Andre drive to arrive at his destination?

____

26 **MARKET** Stacey bought 1.5 pounds of honey ham from the deli. It cost $4.79 per pound. How much did Stacey pay for the honey ham? Do not round.

____

27 **COOKING** A recipe uses 3.5 cups of flour. If Samuel is making 1.5 batches, how many cups of flour will be needed?

____

Vocabulary Check **Write the vocabulary word that completes each sentence.**

28 A(n) _______________ is the answer to a multiplication problem.

29 The numbers that are multiplied together in a multiplication problem are _______________.

30 The _______________ separates the ones place from the tenths place.

31 **Reflect** Using an example, explain how to determine the location of the decimal point in the product.

____

____

____

STOP

Lesson 2-6 Divide Decimals

KEY Concept

Division can be presented three ways.

$\text{divisor}\overline{)\text{dividend}}$ with quotient above	$6\overline{)2.46}$ with 0.41 above
dividend ÷ divisor = quotient	$2.46 \div 6 = 0.41$
$\frac{\text{dividend}}{\text{divisor}} = \text{quotient}$	$\frac{2.46}{6} = 0.41$

To divide a decimal by a decimal, change the divisor to a whole number by moving the decimal to the right. You must move the decimal point to the right the same number of places in the dividend.

$2.2\overline{)3.52}$ becomes $22.\overline{)35.2}$

Move the decimals to the right one place so 2.2 is 22. and 3.52 is 35.2

The decimal point moves straight up into the quotient.

$$\begin{array}{r} 01.6 \\ 22\overline{)35.2} \\ -220 \\ \hline 132 \\ -132 \\ \hline 0 \end{array}$$

Add a zero to keep the place values lined up.

VOCABULARY

dividend
a number that is being divided

divisor
the number you divide the dividend by

decimal point
point separating the ones and tenths in a number

quotient
the answer in a division problem

Always move the decimal place in both the divisor and dividend the same number of places.

Example 1

Find 0.12 ÷ 0.03.

1. Write in vertical form. $0.03\overline{)0.12}$
2. Move the decimal point two places to the right in the divisor and the dividend.

$$\begin{array}{r} 4. \\ 3.\overline{)12.} \\ -12 \\ \hline 0 \end{array}$$

3. Find the quotient.
   0.12 ÷ 0.03 = 4

YOUR TURN!

Find 0.63 ÷ 0.09.

1. Write in vertical form. $\overline{)\quad}$
2. Move the decimal point ________ place(s) to the right in the divisor and the dividend.
3. Find the quotient.
   0.63 ÷ 0.09 = ________

GO ON

Example 2

Find 1.92 ÷ 0.6.

1. To change 0.6 to a whole number, move the decimal to the right one place.

   0.6 becomes 6.

2. Move the decimal one place to the right in the dividend.

   1.92 becomes 19.2

$$\begin{array}{r} 3.2 \\ 6.\overline{)19.2} \\ \underline{-180} \\ 12 \\ \underline{-12} \\ 0 \end{array}$$

3. Divide. Add zeros as needed to line up place values.

4. Move the decimal point straight up from the dividend to the quotient.

   1.92 ÷ 0.6 = 3.2

YOUR TURN!

Find 0.18 ÷ 0.15.

1. To change ________ to a whole number, move the decimal to the right ________ place(s).

   0.15 becomes ________.

2. Move the decimal ________ to the right in the dividend.

   0.18 becomes ________

3. Divide. Add ________ as needed to line up place values.

4. Move the decimal point straight up from the dividend to the quotient.

   0.18 ÷ 0.15 = ________

Guided Practice

Find each quotient.

1. 61.2 ÷ 1.8 = ________

   $\overline{)612.}$

2. 77.7 ÷ 2.1 = ________

   $21.\overline{)777}$

3. 2.6 ÷ 1.3 = ________

4. 293.85 ÷ 4.5 = ________

Step by Step Practice

5 **Find 37.8 ÷ 0.6.**

Step 1 To change ______ to a whole number, move the decimal point to the right ______ place(s).

Step 2 Move the decimal point ______ place(s) to the right in the dividend. 37.8 becomes ______.

Step 3 Divide. Add zeros as needed to line up place values.

Step 4 Move the decimal straight up from the dividend to the quotient.

$06.\overline{)378.}$

$37.8 \div 0.6 =$ ______

Find each quotient.

6 $9.36 \div 1.8 =$ ______

7 $14.4 \div 0.6 =$ ______

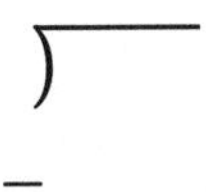

Step by Step Problem-Solving Practice

Solve.

8 **SHOPPING** Wyatt has $9.15 in his wallet. He needs to buy tomatoes that are priced $0.79 each. How many tomatoes can Wyatt buy?

Division can stop when you have a whole number because Wyatt can only buy whole tomatoes.

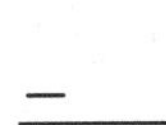

Check off each step.

______ **Understand: I underlined key words.**

______ **Plan: To solve the problem, I will** ______.

______ **Solve: The answer is** ______.

______ **Check: I checked my answer by** ______.

GO ON

Skills, Concepts, and Problem Solving

Find each quotient.

9 $5.6 \div 0.8 =$ ______

10 $7.4 \div 2 =$ ______

11 $8.7 \div 0.3 =$ ______

12 $3.3 \div 5 =$ ______

13 $6.48 \div 0.2 =$ ______

14 $6.48 \div 0.9 =$ ______

15 $8.68 \div 1.4 =$ ______

16 $15.04 \div 1.6 =$ ______

17 $26.3 \div 0.4 =$ ______

Solve.

18 **UNIT PRICE** Hector bought a large popcorn at the movies. What was his price per ounce? Round to the nearest cent.

____

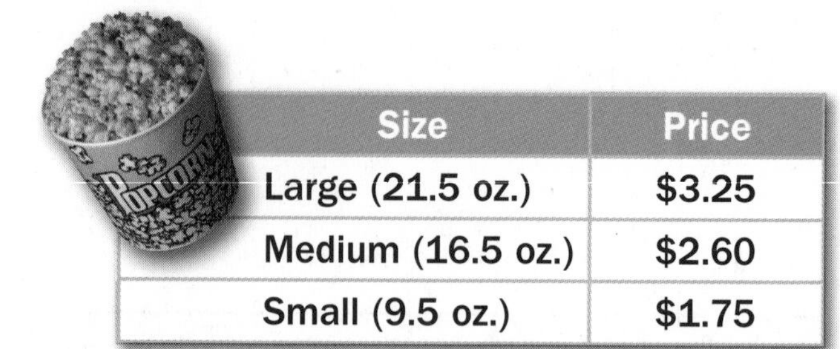

Size	Price
Large (21.5 oz.)	$3.25
Medium (16.5 oz.)	$2.60
Small (9.5 oz.)	$1.75

19 **UNIT PRICE** Tanya bought a small popcorn at the movies. What was her price per ounce? Round to the nearest cent.

____

Vocabulary Check **Write the vocabulary word that completes each sentence.**

20 The answer to a division problem is a ______________________.

21 A number that is being divided is the ______________________.

22 The number you divide the dividend by is the ______________________.

23 **Reflect** How do you use inverse operations to check answers?

____

____

____

STOP

Chapter 2

Progress Check 3 (Lessons 2-5 and 2-6)

Find each product.

1. $0.4 \cdot 0.7 =$ ________

2. $0.2 \cdot 0.9 =$ ________

3. $0.6 \cdot 2.2 =$ ________

4. $4.9 \cdot 0.5 =$ ________

5. $8.3 \cdot 3.8 =$ ________

6. $7.1 \cdot 5.7 =$ ________

7. $\begin{array}{r} 9.02 \\ \times\ 0.6 \\ \hline \end{array}$

8. $\begin{array}{r} 6.12 \\ \times\ 7.4 \\ \hline \end{array}$

9. $\begin{array}{r} 7.56 \\ \times\ 2.08 \\ \hline \end{array}$

Find each quotient.

10. $4.8 \div 0.4 =$ ________

11. $23.8 \div 7 =$ ________

12. $4.88 \div 0.8 =$ ________

13. $2.99 \div 13 =$ ________

14. $56.3 \div 0.5 =$ ________

15. $9.79 \div 1.1 =$ ________

16. $4.8\overline{)18.72}$

17. $7.2\overline{)48.96}$

18. $0.21\overline{)1.0542}$

Solve.

19. **COOKING** Gary is making a casserole for a family reunion. The recipe requires 1.2 pounds of turkey. He is making 3.5 times the amount one recipe makes. How much turkey does he need?

____

20. **READING** Alisa reads at a rate of 1.75 pages per minute. Her next book to read has 875 pages. How many minutes will Alisa need to read this entire book?

____

Chapter 2

Chapter Test

Write each number in the place-value chart.

1 304.073

1000 thousands	100 hundreds	10 tens	1 ones	0.1 tenths	0.01 hundredths	0.001 thousandths	0.0001 ten-thousandths
			.				

2 3.5829

1000 thousands	100 hundreds	10 tens	1 ones	0.1 tenths	0.01 hundredths	0.001 thousandths	0.0001 ten-thousandths
			.				

Use >, <, or = to compare the numbers.

3 8.23 ◯ –8.1

4 –35.2 ◯ –36

5 0.378 ◯ 0.825

6 101.08 ◯ 101

7 0.036 ◯ –0.0356

8 1.882 ◯ 1.228

Round each number to the place value of the underlined digit.

9 35.<u>8</u>27 ________

10 5,7<u>2</u>9.087 ________

11 5.39<u>0</u>4 ________

12 0.9<u>8</u>7 ________

Round 155.09271 to each place value shown.

13 tenth ________

14 hundred ________

15 ten ________

16 one ________

17 hundredth ________

18 thousandth ________

Find each sum or difference.

19 304.22 – 12.899 = ________

20 943.2 + 782.44 = ________

21 290.458 – 5.67 = ________

22 589.05 + 561.2 = ________

Find each product.

23 $0.2 \cdot 0.5 =$ ________

24 $0.8 \cdot 0.9 =$ ________

25 $0.7 \cdot 3.6 =$ ________

26 $8.1 \cdot 0.7 =$ ________

27 $6.4 \cdot 4.4 =$ ________

28 $1.03 \cdot 6.8 =$ ________

29 $\begin{array}{r} 8.17 \\ \times\ 4.1 \\ \hline \end{array}$

30 $\begin{array}{r} 89.3 \\ \times\ 7.3 \\ \hline \end{array}$

31 $\begin{array}{r} 9.09 \\ \times\ 6.12 \\ \hline \end{array}$

Find each quotient.

32 $82.8 \div 23 =$ ________

33 $161.92 \div 0.8 =$ ________

34 $82.8 \div 7.2 =$ ________

35 $23.764 \div 9.14 =$ ________

36 $17.64 \div 8.4 =$ ________

37 $1.16 \div 0.16 =$ ________

Solve.

38 **PACKAGING** Chelsea has 117 pounds of apples to pack into baskets. Each basket needs to weigh 6.5 pounds. How many baskets can Chelsea pack full of apples?

____

39 **CHECK WRITING** When you write a check, the amount of the check has to be written in words. Write the amount \$1,089 in words.

____

Correct the Mistake.

40 Reginald multiplied 502.34 and 2.87 to get a product of 14,417.158. Is his product correct? If not, find and correct his mistake.

____

____

____

STOP

Chapter 3

Fractions and Mixed Numbers

Food and fractions go together.

There are many measuring utensils in the kitchen that are labeled with fractions and mixed numbers. If a recipe calls for $2\frac{1}{3}$ cups of flour, it is important to know the fraction so that the correct utensil is used.

STEP 1 Chapter Pretest Are you ready for Chapter 3? Take the Chapter 3 Pretest to find out.

STEP 2 Preview Get ready for Chapter 3. Review these skills and compare them with what you will learn in this chapter.

What You Know	What You Will Learn
You know how to tell whether a number is divisible by 2, 3, or 5. **Example:** 256 2: Yes. The ones digit, 6, is divisible by 2. 3: No. The sum of the digits, 13, is not divisible by 3. 5: No. The ones digit is neither 0 nor 5. **TRY IT!** 1 Tell if 60 can be divided evenly by 2, 3, or 5. ______________	*Lesson 3-1* To find the **greatest common factor (GCF)** of two or more numbers, you can make a list of common factors. **Example:** 48 and 64 48: 1, 2, 3, 4, 6, 8, 12, (16), 24, 48 64: 1, 2, 4, 8, (16), 32, 64 So, the GCF of 48 and 64 is 16.
You know that $\frac{1}{3}$ names the shaded part of the whole. 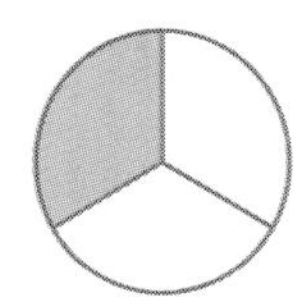 **TRY IT!** **Write the fraction in the simplest form that names the shaded part of the whole.** 2 $\frac{\square}{\square}$ 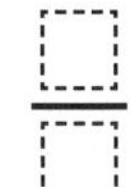  3 $\frac{\square}{\square} = \frac{\square}{\square}$ 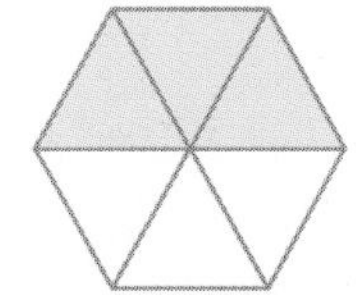	*Lesson 3-3* You can use a model to write $\frac{5}{3}$ as a mixed number. 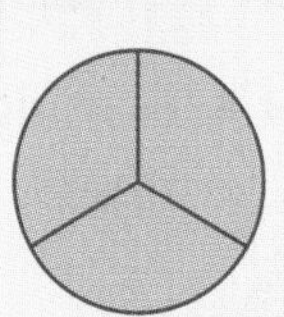 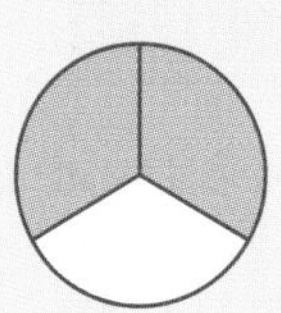 So, $\frac{5}{3} = 1\frac{2}{3}$.

Lesson 3-1

Greatest Common Factors

KEY Concept

You can use lists or factor trees to find the **greatest common factor (GCF)** of two or more whole numbers. Here are two ways to find the GCF of 24 and 32.

List the factors of 24. 1, 2, 3, 4, 6, (8,) 12, 24

List the factors of 32. 1, 2, 4, (8,) 16, 32

8 is the GCF of 24 and 32.

When you make factor trees, use prime factorization.

You can begin with any 2 factors, but use prime factors if you can.

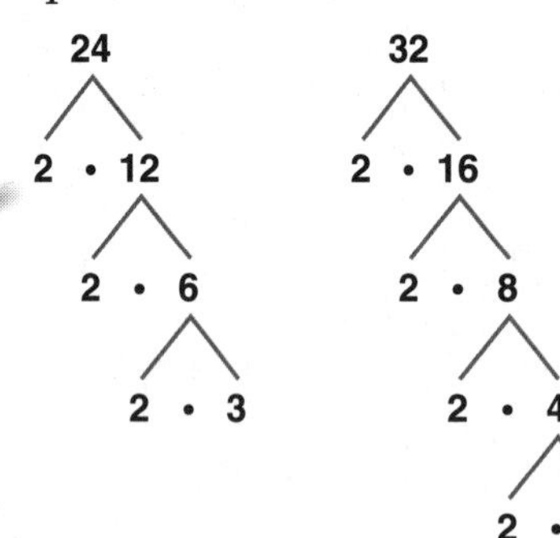

The prime factorization of 24 is $2 \cdot 2 \cdot 2 \cdot 3$.

The prime factorization of 32 is $2 \cdot 2 \cdot 2 \cdot 2 \cdot 2$.

The GCF of 24 and 32 is the product of their common prime factors.

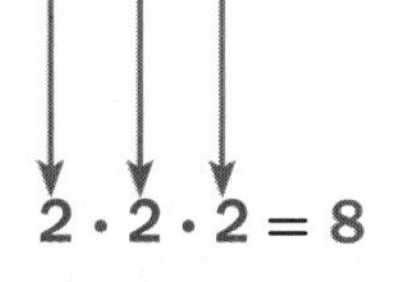

$2 \cdot 2 \cdot 2 = 8$

VOCABULARY

factors
the quantities being multiplied

greatest common factor (GCF)
the product of the prime factors common to two or more integers

prime factorization
a whole number expressed as a product of factors that are all prime numbers

prime number
a whole number greater than 1 with factors that are only 1 and itself

You can find the GCF of algebraic expressions using the same methods as you do whole numbers.

Example 1

Find the GCF of 28 and 16 by listing factors.

1. List the factors of 28 and 16.

   28: 1, 2, 4, 7, 14, 28

   16: 1, 2, 4, 8, 16

2. Name the common factors.

   1, 2, 4

3. The GCF is 4.

YOUR TURN!

Find the GCF of 24 and 40 by listing factors.

1. List the factors of 24.

   24: ____________________

   40: ____________________

2. Name the common factors.

   ____

3. The GCF is ______.

Example 2

Find the GCF of 30 and 54 by using factor trees.

1. Make factor trees for each number.
2. The prime factorization of 30 is $2 \cdot 3 \cdot 5$.

   The prime factorization of 54 is $2 \cdot 3 \cdot 3 \cdot 3$.
3. Find the product of their common prime factors.

   $2 \cdot 3 = 6$

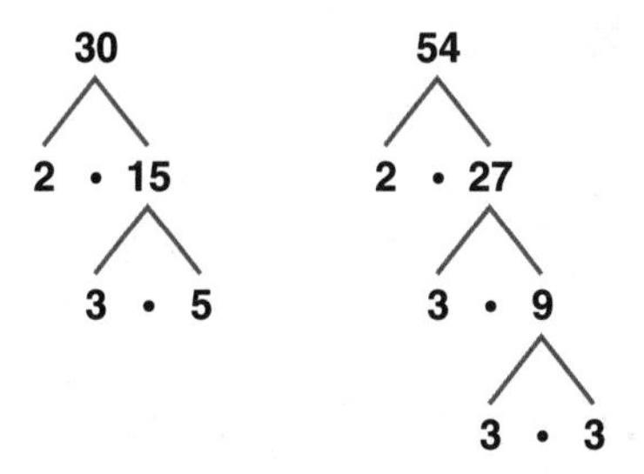

YOUR TURN!

Find the GCF of 20 and 28 by using factor trees.

1. Make factor trees for each number.
2. The ________________ of 20 is ________________.

   The ________________ of 28 is ________________.
3. Find the product of their common prime factors.

   _______ • _______ = _______

20

28

Example 3

Find the GCF of x^2y^4 and xy^2.

1. Write each expression as a product of prime factors.

   $x^2y^4 = x \cdot x \cdot y \cdot y \cdot y \cdot y$

   $xy^2 = x \cdot y \cdot y$
2. Circle the common factors.
3. Find the product of their common factors.

   xy^2

YOUR TURN!

Find the GCF of x^3y and x^4y^2.

1. Write each expression as a product of prime factors.

   x^3y = ________________

   x^4y^2 = ________________
2. Circle the common factors.
3. Find the product of their common factors.

   ____

GO ON

Guided Practice

Find the GCF of each set of numbers by listing factors.

1. 56 and 70 ______

   56: ______

   70: ______

2. 26 and 72 ______

   26: ______

   72: ______

Find the GCF of each set of numbers by using factor trees.

3. 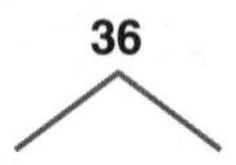 

   36: ______

   44: ______

   ______ • ______ = ______

   The GCF is ______.

4.  

   20: ______

   70: ______

   ______ • ______ = ______

   The GCF is ______.

Step by Step Practice

5. Find the GCF of xy^2 and y^3.

   Step 1 Factor each expression.
   xy^2 = ______
   y^3 = ______

   Step 2 Circle the common factors.

   Step 3 Find the product of their common factors. ______ • ______ = ______

Find the GCF of these pairs of expressions.

6. a^2bc and a^3c^2 ______

   a^2bc = ______

   a^3c^2 = ______

7. x^3y and y^2z^2 ______

   x^3y = ______

   y^2z^2 = ______

Step by Step Problem-Solving Practice

Solve.

8 **SCHOOL** Seventy-six fourth graders and 84 fifth graders enter a gymnasium for a school assembly. If each row of seats has the same number of students, what is the greatest number of students per row?

Multiples of 76: ______________________________

Multiples of 84: ______________________________

Check off each step.

______ **Understand: I underlined key words.**

______ **Plan: To solve the problem, I will** ______________________________.

______ **Solve: The answer is** ______________________________.

______ **Check: I checked my answer by** ______________________________.

Skills, Concepts, and Problem Solving

Find the GCF of each set of numbers.

9 12, 72 ______

10 50, 60 ______

11 33, 44 ______

12 27, 81 ______

13 40, 125 ______

14 68, 84 ______

15 18, 24, 30 ______

16 35, 42, 84 ______

17 36, 60, 100 ______

18 9, 27, 77 ______

19 24, 72, 108 ______

20 12, 30, 114 ______

GO ON

Find the GCF of each set of numbers.

21 x^2y^5, x^4y^2 ______

22 x^3y^2, xy^5 ______

23 x^4y^6, xy^3 ______

24 xy^6, x^5y^5 ______

25 x^3y^6, x^5y^2 ______

26 x^5y^3, x^6y ______

Solve.

FIELD TRIP Mrs. Gordon's class is going on a field trip to the city zoo. She collected money each day as shown in the table. Each student paid the same amount of money.

Class Field Trip Collection	
Monday	$104
Tuesday	$52
Wednesday	$78
Thursday	$91

27 What is the most each student paid to go on the field trip?

____

28 What is the least number of students who have paid to go on the field trip?

____

Vocabulary Check **Write the vocabulary word that completes each sentence.**

29 The ______________________ is the product of the prime factors common to two or more integers.

30 Expressing a composite number as a product of its prime factors is known as ______________________.

31 Two numbers multiplied together to get a product are ______________________.

32 A(n) ______________________ is a whole number with exactly two factors, 1 and itself.

33 **Reflect** Determine if the following statement is *true, false,* or *sometimes true.* Give an example.

The GCF of two numbers is less than both numbers.

____

____

____

____

STOP

Lesson 3-2

Simplify Fractions

KEY Concept

A fraction expressed in **simplest form** is equivalent to the original fraction.

$$\frac{6}{18} = \frac{2}{6} = \frac{1}{3}$$

$\frac{1}{3}$ is in simplest form.

One way to express a fraction in simplest form is to divide the **numerator** and **denominator** by their **GCF** to express a fraction in simplest form.

$$\frac{3 \div 3}{12 \div 3} = \frac{1}{4}$$

Another way is to write the numerator and denominator as products of prime factors. Then you can reduce common factors.

$$\frac{12}{16} = \frac{\cancel{2} \cdot \cancel{2} \cdot 3}{\cancel{2} \cdot \cancel{2} \cdot 2 \cdot 2} = \frac{3}{2 \cdot 2} = \frac{3}{4}$$

VOCABULARY

denominator
the number below the bar in a fraction that tells how many equal parts are in the whole or the set

equivalent fractions
fractions that represent the same number

greatest common factor (GCF)
the product of the prime factors common to two or more integers

numerator
the number above the bar in a fraction that tells how many equal parts are being used

simplest form
the form of a fraction when the GCF of the numerator and denominator is 1

A fraction is not in simplest form if the numerator and denominator have any common factors. Continue to divide the numerator and denominator until there are no more common factors. Improper fractions must be written as mixed numbers to be in simplest form.

Example 1

Write $\frac{10}{35}$ in simplest form.

1. Find the GCF by writing all factors of each number.

   10: 1, 2, (5,) 10

   35: 1, (5,) 7, 35

2. The GCF is 5.

3. Divide the numerator and denominator by 5.

   $$\frac{10 \div 5}{35 \div 5} = \frac{2}{7}$$

YOUR TURN!

Write $\frac{16}{20}$ in simplest form.

1. Find the ______ by writing all factors of each number.

   16: ______________________

   20: ______________________

2. The GCF is ______.

3. Divide the numerator and denominator by ______.

   $$\frac{16 \div \square}{20 \div \square} = \frac{\square}{\square}$$

GO ON

Example 2

Write $\frac{35}{175}$ in simplest form.

1. Write the numerator and denominator as a product of prime factors.

$$\frac{35}{175} = \frac{5 \cdot 7}{5 \cdot 5 \cdot 7}$$

2. Reduce common factors.

$$\frac{35}{175} = \frac{\cancel{5} \cdot \cancel{7}}{\cancel{5} \cdot 5 \cdot \cancel{7}} = \frac{1}{5}$$

YOUR TURN!

Write $\frac{56}{196}$ in simplest form.

1. Write the numerator and denominator as a product of prime factors.

$$\frac{56}{196} = \frac{\square \cdot \square \cdot \square \cdot \square}{\square \cdot \square \cdot \square \cdot \square}$$

2. Reduce common factors.

$$\frac{56}{196} = \frac{\square \cdot \square \cdot \square \cdot \square}{\square \cdot \square \cdot \square \cdot \square} = \frac{\square}{\square}$$

Example 3

Write $\frac{6x^2y^3}{8xy^2}$ in simplest form.

1. Factor each expression.

$$\frac{6x^2y^3}{8xy^2} = \frac{2 \cdot 3 \cdot x \cdot x \cdot y \cdot y \cdot y}{2 \cdot 2 \cdot 2 \cdot x \cdot y \cdot y}$$

2. Reduce common factors.

$$\frac{6x^2y^3}{8xy^2} = \frac{\cancel{2} \cdot 3 \cdot \cancel{x} \cdot x \cdot \cancel{y} \cdot \cancel{y} \cdot y}{\cancel{2} \cdot 2 \cdot 2 \cdot \cancel{x} \cdot \cancel{y} \cdot \cancel{y}} = \frac{3xy}{4}$$

YOUR TURN!

Write $\frac{12y}{20x^3y^2}$ in simplest form.

1. Factor each expression.

$$\frac{12y}{20x^3y^2} = \frac{\square \cdot \square \cdot \square \cdot \square}{\square \cdot \square \cdot \square \cdot \square \cdot \square \cdot \square \cdot \square \cdot \square}$$

2. Reduce common factors.

$$\frac{12y}{20x^3y^2} = \frac{\square \cdot \square \cdot \square \cdot \square}{\square \cdot \square \cdot \square \cdot \square \cdot \square \cdot \square \cdot \square \cdot \square} = \frac{\square}{\square}$$

Guided Practice

Write each fraction in simplest form.

1. $\frac{10}{45}$

   Factors of 10: ______________________

   Factors of 45: ______________________

   $\frac{10}{45} \div \frac{\square}{\square} = \frac{\square}{\square}$

2. $\frac{14}{56}$

   Factors of 14: ______________________

   Factors of 56: ______________________

   $\frac{14}{56} \div \frac{\square}{\square} = \frac{\square}{\square}$

Write each fraction in simplest form.

3. $\frac{12}{21} = \frac{\square \cdot \square \cdot \square}{\square \cdot \square} = \frac{\square}{\square}$

4. $\frac{48}{64} = \frac{\square \cdot \square \cdot \square \cdot \square \cdot \square}{\square \cdot \square \cdot \square \cdot \square \cdot \square \cdot \square} = \frac{\square}{\square}$

Step by Step Practice

5. Write $\frac{9a^2b}{21a^2c^2}$ in simplest form.

   Step 1 Factor each expression.

   $$\frac{9a^2b}{21a^2c^2} = \frac{\square \cdot \square \cdot \square \cdot \square \cdot \square}{\square \cdot \square \cdot \square \cdot \square \cdot \square \cdot \square}$$

   Step 2 Reduce the common prime factors.

   $$\frac{9a^2b}{21a^2c^2} = \frac{\square \cdot \square \cdot \square \cdot \square \cdot \square}{\square \cdot \square \cdot \square \cdot \square \cdot \square \cdot \square} = \frac{\square}{\square}$$

Write each fraction in simplest form.

6. $\frac{6xy^4}{28x^2y^2} = \frac{\square \cdot \square \cdot \square \cdot \square \cdot \square \cdot \square \cdot \square}{\square \cdot \square \cdot \square \cdot \square \cdot \square \cdot \square \cdot \square} = \frac{\square}{\square}$

7. $\frac{10xz^2}{34x^3yz} = \frac{\square \cdot \square \cdot \square \cdot \square \cdot \square}{\square \cdot \square \cdot \square \cdot \square \cdot \square \cdot \square \cdot \square} = \frac{\square}{\square}$

8. $\frac{22a^4b}{26a^6b^5} =$ __________

9. $\frac{10m^8n^3}{16m^6n^6} =$ __________

GO ON

Step by Step Problem-Solving Practice

Solve.

10 **SLEEP** Rosina spends 8 hours a day sleeping. In simplest form, what fraction of a week does Hope spend sleeping?

7 days • _____ hours = _____ hours a week sleeping

7 days • _____ hours = _____ total hours a week

Write all factors.

56: _____ • _____ • _____ • _____

168: _____ • _____ • _____ • _____ • _____

GCF _____ • _____ • _____ • _____ = _____.

$\frac{\square \text{ hours sleeping}}{\square \text{ total hours}} \div \frac{\square}{\square} = \frac{\square}{\square}$

Check off each step.

_____ **Understand: I underlined key words.**

_____ **Plan: To solve the problem, I will** _____________________.

_____ **Solve: The answer is** _____________________.

_____ **Check: I checked my answer by** _____________________.

Skills, Concepts, and Problem Solving

Write each fraction in simplest form.

11 $\frac{5}{20} = \frac{\square}{\square}$

12 $\frac{28}{35} = \frac{\square}{\square}$

13 $\frac{20}{28} = \frac{\square}{\square}$

14 $\frac{16}{120} = \frac{\square}{\square}$

15 $\frac{12}{33} = \frac{\square}{\square}$

16 $\frac{60}{85} = \frac{\square}{\square}$

Write each fraction in simplest form.

17 $\frac{21x^4}{49x^3} =$ ________

18 $\frac{2z^2}{40z^6} =$ ________

19 $\frac{16m^3n^4}{6mn^5} =$ ________

20 $\frac{9cd^4}{27c^6d^5} =$ ________

21 $\frac{24y^4}{42y^4} =$ ________

22 $\frac{32n^4p^8}{38n^2p^3} =$ ________

Solve. Write the answer in simplest form.

23 **TIME** Twenty seconds is what part of one minute?

____

24 **MEASUREMENT** Eighty feet is what part of a mile? (Hint: There are 5,280 feet in one mile.)

____

Vocabulary Check **Write the vocabulary word that completes each sentence.**

25 Fractions that represent the same number are ________________.

26 A fraction in which the numerator and denominator have no common factor greater than 1 is written in ________________.

27 The product of the prime factors common to two or more integers is the ________________.

28 **Reflect** Two students wrote the fraction $\frac{18}{42}$ in simplest form. Who is correct? Explain your answer.

Nikki writes $\frac{18}{42} \div \frac{2}{2} = \frac{9}{21}$. Brett writes $\frac{18}{42} \div \frac{6}{6} = \frac{3}{7}$.

____

____

____

____

STOP

Progress Check 1 (Lessons 3-1 and 3-2)

Find the GCF of each set of numbers.

1. 14, 42 ______
2. 30, 65 ______
3. 39, 65 ______
4. 63, 72 ______
5. 56, 104 ______
6. 28, 84 ______
7. 9, 24, 33 ______
8. 60, 85, 100 ______
9. 55, 110, 121 ______

Write each fraction in simplest form.

10. $\frac{8}{18} = \frac{\square}{\square}$
11. $\frac{21}{35} = \frac{\square}{\square}$
12. $\frac{56}{72} = \frac{\square}{\square}$
13. $\frac{30}{75} = \frac{\square}{\square}$
14. $\frac{10}{45} = \frac{\square}{\square}$
15. $\frac{26}{65} = \frac{\square}{\square}$
16. $\frac{12x^3y^7}{28x^4y} = \frac{\square}{\square}$
17. $\frac{48x^5z^6}{16x^4z^2} = \frac{\square}{\square} = \square$

Solve. Write the answer in simplest form.

18. **ENTERTAINMENT** Groups of students were admitted into a fun house. The grades of the students were recorded in the table at the right.

The number of students in each group was equal. What is the most number of students that could be in a group?

Visitors to the Fun House	
Eighth Grade	63
Ninth Grade	35
Tenth Grade	42
Eleventh Grade	35

____

19. **LENGTH** The side of a dog house is one yard long. Its door is 18 inches wide. What part of a side of the house is the door? (Hint: There are 36 inches in one yard.)

____

20. **RECIPE** Rico used 76 ounces of juice to make punch. A gallon is 128 ounces. What part of a gallon did Rico use?

____

Lesson 3-3

Mixed Numbers

KEY Concept

Fractions greater than 1 can be written as mixed numbers or as improper fractions.

Mixed Numbers

$$2\frac{3}{5} = 2 \text{ wholes} + \frac{3}{5}$$

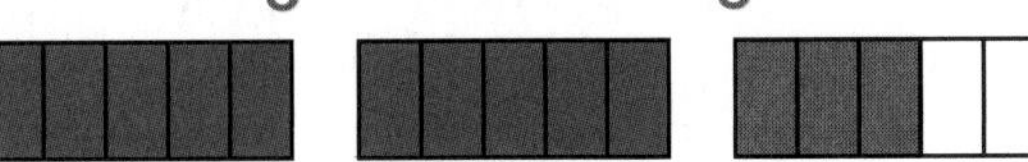

Improper Fractions

$$\frac{13}{8}$$

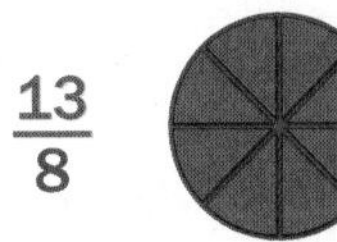
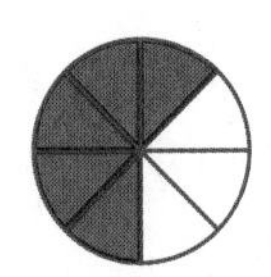

To change an improper fraction to a mixed number, divide the numerator by the denominator.

$$\frac{13}{8} = 8\overline{)13} = 1\frac{5}{8}$$

$$\begin{array}{r} 1 \\ 8\overline{)13} \\ -8 \\ \hline 5 \end{array}$$

The quotient is the whole number part. The remainder is the numerator. The divisor is the denominator.

To change a mixed number to an improper fraction, multiply the whole number by the denominator and add the product to the numerator.

$$1\frac{5}{8} = \frac{(1 \times 8) + 5}{8} = \frac{13}{8}$$

The numerator is the product of the whole number and the denominator, plus the numerator. The denominator does not change.

VOCABULARY

denominator
the number below the bar in a fraction that tells how many equal parts are in the whole or the set

improper fraction
a fraction with a numerator that is greater than or equal to the denominator

mixed number
a number that has a whole number part and a fraction part

numerator
the number above the bar in a fraction that tells how many equal parts are being used

Example 1

Use a model to write $\frac{7}{5}$ as a mixed number.

1. Draw a model to represent the fraction. Each whole is divided into 5 parts. Shade 7 of the parts.

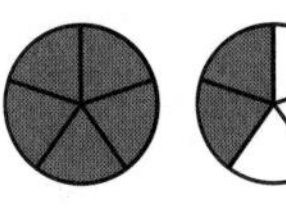

2. Write the whole number and the fraction.

$1\frac{2}{5}$

YOUR TURN!

Use a model to write $\frac{5}{2}$ as a mixed number.

1. Draw a model to represent the fraction. Each whole is divided into _____ parts. Shade _____ of the parts.

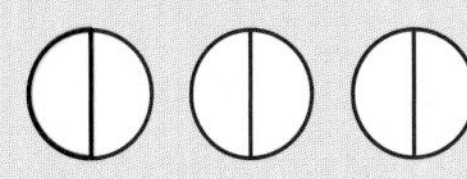

2. Write the whole number and the fraction.

_____

GO ON

Example 2

Write $\frac{11}{6}$ as a mixed number.

1. Divide the numerator by the denominator.

$$\begin{array}{r} 1 \\ 6\overline{)11} \\ -6 \\ \hline 5 \end{array}$$

2. Write the quotient as the whole number. Write the remainder over the denominator.

$1\frac{5}{6}$

YOUR TURN!

Write $\frac{14}{4}$ as a mixed number.

1. Divide the numerator by the denominator. $4\overline{)14}$ $-$ _____

2. Write the quotient as the whole number. Write the remainder over the denominator.

_____ = _____

Example 3

Write $3\frac{4}{5}$ as an improper fraction.

1. Multiply the whole number by the denominator.

$5 \cdot 3 = 15$

2. Add the numerator.

$15 + 4 = 19$

3. Write the sum over the denominator.

$\frac{19}{5}$

YOUR TURN!

Write $1\frac{2}{9}$ as an improper fraction.

1. Multiply the whole number by the denominator

_____ • _____ = _____

2. Add the numerator.

_____ + _____ = _____

3. Write the sum over the denominator.

_____

Guided Practice

Write each improper fraction as a mixed number in simplest form.

1. $\frac{9}{4}$ = _____

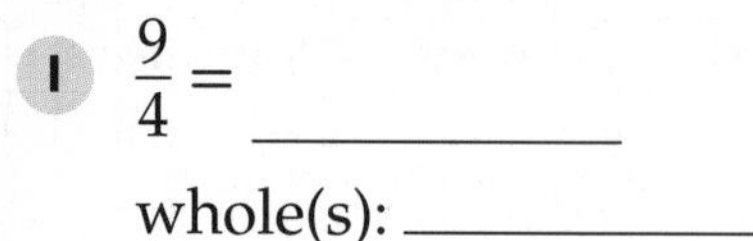
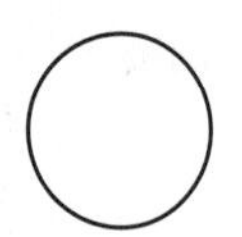
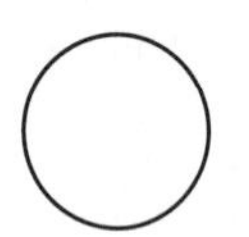
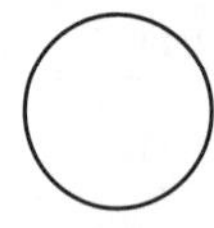

whole(s): _____

fraction: _____

2. $\frac{7}{5}$ = _____ $5\overline{)7}$ $-$ _____

3. $\frac{20}{6}$ = _____ $6\overline{)20}$ $-$ _____

Step by Step Practice

4 Write $3\frac{2}{5}$ as an improper fraction.

Step 1 Multiply the whole number by the denominator.

_____ • _____ = _____

Step 2 Add the numerator.

_____ + _____ = _____

Step 3 Write the sum over the denominator.

_____

Write each mixed number as an improper fraction.

5 $2\frac{1}{3}$ = _____

_____ • _____ = _____

_____ + _____ = _____

6 $4\frac{2}{5}$ = _____

_____ • _____ = _____

_____ + _____ = _____

Step by Step Problem-Solving Practice

Solve.

7 **BAGELS** Mrs. Sato bought 78 bagels for the eighth grade class on their last day of school. How many dozen bagels did Mrs. Sato buy? Write your answer as a mixed number in simplest form.

$12\overline{)78}$

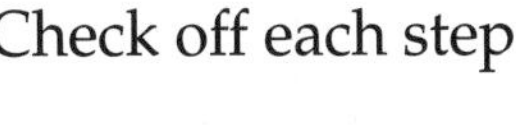

Check off each step.

_____ **Understand: I underlined key words.**

_____ **Plan: To solve the problem, I will** ____________________.

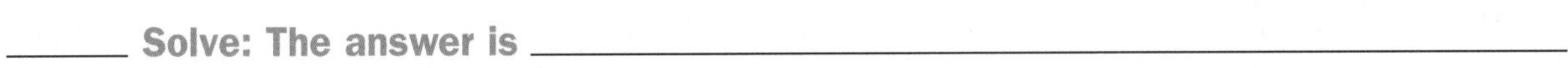

_____ **Solve: The answer is** ____________________.

_____ **Check: I checked my answer by** ____________________.

GO ON

Skills, Concepts, and Problem Solving

Write each improper fraction as a mixed number in simplest form.

8. $\frac{7}{3} =$ ______

9. $\frac{25}{6} =$ ______

10. $\frac{68}{6} =$ ______

11. $\frac{48}{11} =$ ______

12. $\frac{81}{2} =$ ______

13. $\frac{89}{10} =$ ______

14. $\frac{105}{20} =$ ______

15. $\frac{29}{3} =$ ______

16. $\frac{100}{8} =$ ______

Solve.

17. **CARPENTRY** James is building a sandbox and needs $21\frac{3}{8}$ feet of wood. How much wood does James need? Write your answer as an improper fraction.

______

18. **CROSS COUNTRY** Winona ran $\frac{37}{8}$ miles in her cross country meet before injuring her foot. How many miles did Winona run? Write your answer as a mixed number in simplest form.

______

Vocabulary Check **Write the vocabulary word that completes each sentence.**

19. A fraction with a numerator that is greater than or equal to the denominator is called a(n) ______.

20. A(n) ______ is a number that has a whole number part and a fraction part.

21. **Reflect** How do you know when a fraction can be written as a mixed number?

______

______

______

STOP

Lesson 3-4 Compare and Order Fractions

KEY Concept

You can use a number line to compare fractions. On a number line, the greater fraction is to the right. $\frac{4}{3}$ is greater than $\frac{2}{3}$.

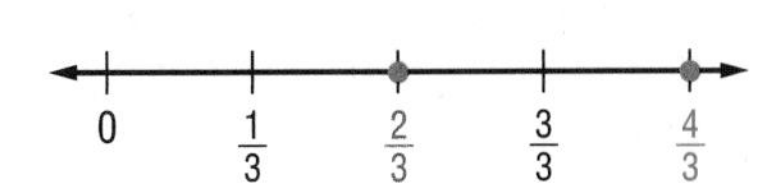

$\frac{4}{3} > \frac{2}{3}$

Compare fractions with like denominators by comparing the numerators.

To compare fractions without using a number line, write all fractions so that they have common denominators. Then, compare their numerators.

$\frac{8 \cdot 3}{6 \cdot 3} = \frac{24}{18}$ $\quad \frac{11 \cdot 2}{9 \cdot 2} = \frac{22}{18}$

The LCD of 6 and 9 is 18. Write both fractions with 18 as the denominator.

Because $24 > 22$, $\frac{24}{18} > \frac{22}{18}$.

So, $\frac{8}{6} > \frac{11}{9}$.

VOCABULARY

least common denominators (LCD)
the least common multiple of the denominators of 2 or more fractions

least common multiple (LCM)
the least whole number greater than one that is a common multiple of each of two or more numbers

multiple
the product of a number and any whole number

To compare mixed numbers, rewrite them as improper fractions and then compare.

Example 1

Write >, <, or = to compare $\frac{6}{4}$ and $\frac{9}{4}$. Use a number line.

1. Graph both numbers on a number line.

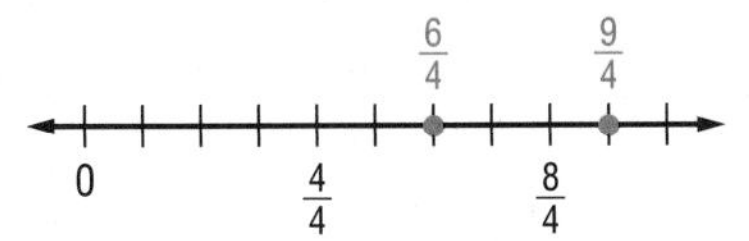

2. Because $\frac{9}{4}$ is to the right of $\frac{6}{4}$, it is the greater number.
3. Write an inequality. $\frac{6}{4} < \frac{9}{4}$

YOUR TURN!

Write >, <, or = to compare $\frac{4}{3}$ and $\frac{3}{3}$. Use a number line.

1. Graph both numbers on a number line.

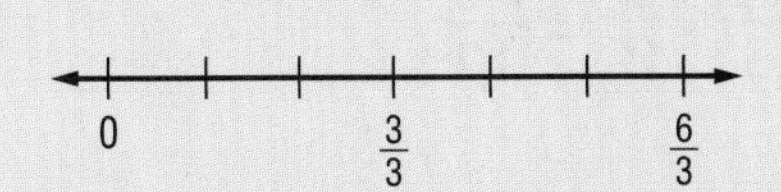

2. Because $\frac{4}{3}$ is to the ________ of $\frac{3}{3}$, it is the ________ number.
3. Write an inequality. $\frac{4}{3} \bigcirc \frac{3}{3}$

GO ON

Example 2

Write >, <, or = to compare $\frac{5}{2}$ and $\frac{7}{3}$.

1. Find the LCD. The LCD is 6.

 2: 2, 4, 6, 8, 10, . . .

 3: 3, 6, 9, 12, . . .

2. Write both fractions with 6 in the denominator.

 $\frac{5}{2} \cdot \frac{3}{3} = \frac{15}{6}$ $\frac{7}{3} \cdot \frac{2}{2} = \frac{14}{6}$

3. Compare numerators. **15 > 14**
4. Write the inequality. $\frac{5}{2} > \frac{7}{3}$

YOUR TURN!

Write >, <, or = to compare $\frac{7}{4}$ and $\frac{4}{5}$.

1. Find the LCD. The LCD is ______.

 4: ____________________

 5: ____________________

2. Write both fractions with ______ in the denominator.

 $\frac{7}{4} \cdot \frac{\square}{\square} = \frac{\square}{\square}$ $\frac{4}{5} \cdot \frac{\square}{\square} = \frac{\square}{\square}$

3. Compare numerators. ______ ◯ ______
4. Write the inequality. $\frac{7}{4}$ ◯ $\frac{4}{5}$

Example 3

Order the fractions from least to greatest.

$\frac{11}{4}, 1\frac{5}{12}, \frac{13}{6}$

1. Find the LCD. The LCD is 12.

 4: 4, 8, 12, 16

 6: 6, 12, 18, 24

 12: 12, 24, 36

2. Write the fractions with 12 in the denominator. Rewrite mixed numbers as improper fractions.

 $\frac{11}{4} \cdot \frac{3}{3} = \frac{33}{12}$ $\frac{17}{12} \cdot \frac{1}{1} = \frac{17}{12}$

 $\frac{13}{6} \cdot \frac{2}{2} = \frac{26}{12}$

3. Compare numerators to order the fractions.

 $1\frac{5}{12}, \frac{13}{6}, \frac{11}{4}$

YOUR TURN!

Order the fractions from least to greatest.

$1\frac{7}{8}, 3\frac{1}{2}, \frac{17}{6}$

1. Find the LCD. The LCD is ______.

 8: ____________________

 2: ____________________

 6: ____________________

2. Write the fractions with ______ in the denominator. Rewrite mixed numbers as improper fractions.

 $\frac{\square}{8} \cdot \frac{\square}{\square} = \frac{\square}{\square}$ $\frac{\square}{2} \cdot \frac{\square}{\square} = \frac{\square}{\square}$

 $\frac{17}{6} \cdot \frac{\square}{\square} = \frac{\square}{\square}$

3. Compare numerators to order the fractions.

 ______, ______, ______

Guided Practice

Write >, <, or = to compare the fractions.

1. $\frac{6}{7} \bigcirc \frac{8}{7}$

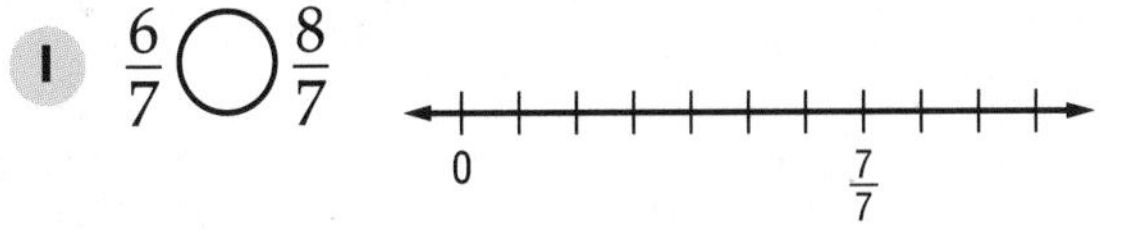

2. $\frac{9}{5} \bigcirc \frac{5}{5}$

0 $\quad \frac{5}{5} \quad \frac{10}{5}$

3. $\frac{5}{6} \bigcirc \frac{7}{10}$

6: ______________________

10: ______________________

The LCD is ______.

$\frac{5}{6} \cdot \frac{\square}{\square} = \frac{\square}{\square}$ $\qquad \frac{7}{10} \cdot \frac{\square}{\square} = \frac{\square}{\square}$

______ $\bigcirc$ ______, so $\frac{\square}{\square} \bigcirc \frac{\square}{\square}$

4. $\frac{3}{5} \bigcirc \frac{7}{8}$

5: ______________________

8: ______________________

The LCD is ______.

$\frac{3}{5} \cdot \frac{\square}{\square} = \frac{\square}{\square}$ $\qquad \frac{7}{8} \cdot \frac{\square}{\square} = \frac{\square}{\square}$

______ $\bigcirc$ ______, so $\frac{\square}{\square} \bigcirc \frac{\square}{\square}$

Step by Step Practice

5. Order the fractions from least to greatest. $\frac{15}{4}, 2\frac{3}{5}, 1\frac{3}{10}$

Step 1 Rewrite mixed numbers as improper fractions.

$2\frac{3}{5} = \frac{\square}{5}$ $\qquad 1\frac{3}{10} = \frac{\square}{10}$

Step 2 Find the LCD.

4: ______________________

5: ______________________

10: ______________________

The LCD is ______.

Step 3 Write the fractions with the LCD in the denominator.

$\frac{15}{4} \cdot \frac{\square}{\square} = \frac{\square}{\square}$ $\qquad \frac{\square}{5} \cdot \frac{\square}{\square} = \frac{\square}{\square}$ $\qquad \frac{13}{10} \cdot \frac{\square}{\square} = \frac{\square}{\square}$

Step 4 Compare numerators to order the fractions. ______________________

GO ON

Order the fractions from least to greatest.

6 $\frac{14}{3}, 1\frac{5}{6}, 3\frac{3}{5}$

$1\frac{5}{6} = \frac{\square}{6}$ $3\frac{3}{5} = \frac{\square}{5}$

3: ______________________

____

5: ______________________

6: ______________________

The LCD is ____.

$\frac{14}{3} \cdot \frac{\square}{\square} = \frac{\square}{\square}$ $\frac{11}{6} \cdot \frac{\square}{\square} = \frac{\square}{\square}$

$\frac{18}{5} \cdot \frac{\square}{\square} = \frac{\square}{\square}$

The order is ____, ____, ____.

7 $2\frac{2}{3}, \frac{20}{7}, \frac{17}{6}$

$2\frac{2}{3} = \frac{\square}{3}$

3: ______________________

____

6: ______________________

7: ______________________

The LCD is ____.

$\frac{8}{3} \cdot \frac{\square}{\square} = \frac{\square}{\square}$ $\frac{20}{7} \cdot \frac{\square}{\square} = \frac{\square}{\square}$

$\frac{17}{6} \cdot \frac{\square}{\square} = \frac{\square}{\square}$

The order is ____, ____, ____.

Step by Step Problem-Solving Practice

Solve.

8 FINANCIAL LITERACY Wesley is trying to decide between buying lemonade or a slushy drink. The lemonade is $12\frac{1}{2}$ ounces and the slushy drink is $\frac{35}{3}$ ounces. Both drinks cost \$1.75. Which drink should Wesley choose if he wants the larger drink?

$\frac{35}{3} =$ ____

$12\frac{1}{2} \bigcirc$ ____

Wesley should choose the __________ lemonade.

Check off each step.

____ Understand: I underlined key words.

____ Plan: To solve the problem, I will ______________________.

____ Solve: The answer is ______________________.

____ Check: I checked my answer by ______________________.

Skills, Concepts, and Problem Solving

Use >, <, or = to compare each pair of fractions.

9. $\frac{1}{5} \bigcirc \frac{1}{4}$
10. $\frac{3}{8} \bigcirc \frac{7}{16}$
11. $\frac{1}{5} \bigcirc \frac{3}{20}$
12. $\frac{5}{8} \bigcirc \frac{30}{48}$
13. $\frac{7}{10} \bigcirc \frac{3}{10}$
14. $\frac{7}{15} \bigcirc \frac{7}{11}$
15. $\frac{4}{7} \bigcirc \frac{5}{8}$
16. $\frac{16}{20} \bigcirc \frac{40}{50}$
17. $\frac{2}{3} \bigcirc \frac{4}{9}$

Solve.

TRACK AND FIELD Ava was in the high jump event at her high school track meet. She was given three attempts. Her results are shown in the table at the right.

Attempt	Distance (ft)
first	$4\frac{5}{6}$ feet
second	$\frac{31}{6}$ feet
third	$\frac{11}{2}$ feet

18. Which attempt was Ava's best jump?

____

19. Order the attempts from least to greatest.

____

Vocabulary Check **Write the vocabulary word that completes each sentence.**

20. The least common multiple of the denominators of two or more fractions is the ______________________.

21. ______________________ are made up of the product of a number and any whole number.

22. Denominators that are not the same are also called ______________________.

23. Denominators that are the same are also called ______________________.

24. **Reflect** Determine which number is closest to 5: $\frac{19}{4}$, $\frac{1}{5}$, or 6.

____

STOP

Chapter 3

Progress Check 2 (Lessons 3-3 and 3-4)

Write each improper fraction as a mixed number in simplest form.

1 $\frac{8}{3} =$ ______

2 $\frac{30}{4} =$ ______

3 $\frac{80}{13} =$ ______

4 $\frac{32}{10} =$ ______

5 $\frac{66}{7} =$ ______

6 $\frac{75}{2} =$ ______

Write each mixed number as an improper fraction.

7 $9\frac{1}{6} =$ ______

8 $7\frac{1}{5} =$ ______

9 $8\frac{2}{11} =$ ______

10 $5\frac{3}{5} =$ ______

11 $10\frac{2}{7} =$ ______

12 $15\frac{3}{4} =$ ______

Write >, <, or = to compare each fraction pair.

13 $\frac{1}{3}$ ◯ $\frac{1}{4}$

14 $\frac{3}{7}$ ◯ $\frac{6}{7}$

15 $\frac{2}{5}$ ◯ $\frac{4}{10}$

16 $\frac{7}{9}$ ◯ $\frac{16}{45}$

17 $\frac{5}{12}$ ◯ $\frac{2}{3}$

18 $\frac{5}{16}$ ◯ $\frac{7}{10}$

Order the fractions from least to greatest.

19 $\frac{15}{12}, 1\frac{1}{6}, 1\frac{3}{10}$ ______________

20 $3\frac{1}{2}, \frac{17}{5}, 2\frac{9}{10}$ ______________

21 $\frac{16}{7}, 1\frac{3}{28}, 2\frac{1}{4}$ ______________

22 $\frac{8}{9}, 1\frac{19}{20}, 1\frac{3}{15}$ ______________

Solve.

23 **RUNNING** Kamal ran a total of $7\frac{4}{5}$ mile. What improper fraction shows the distance Kamal ran?

____

24 **VOLUME** Larry uses $4\frac{7}{8}$ cups of water to fill a goldfish bowl. What improper fraction represents the amount of water Larry used?

____

Lesson 3-5

Multiply Fractions

KEY Concept

Fractions can be multiplied by whole numbers or by other fractions. In either case, you can think of multiplication as repeated addition. For example, $\frac{3}{5} \cdot 3$ is the same as $\frac{3}{5} + \frac{3}{5} + \frac{3}{5}$.

$$\frac{3}{5} \cdot 3 = \frac{3}{5} + \frac{3}{5} + \frac{3}{5} = \frac{9}{5} \text{ or } 1\frac{4}{5}$$

When multiplying a fraction by another fraction, you can use an area model.

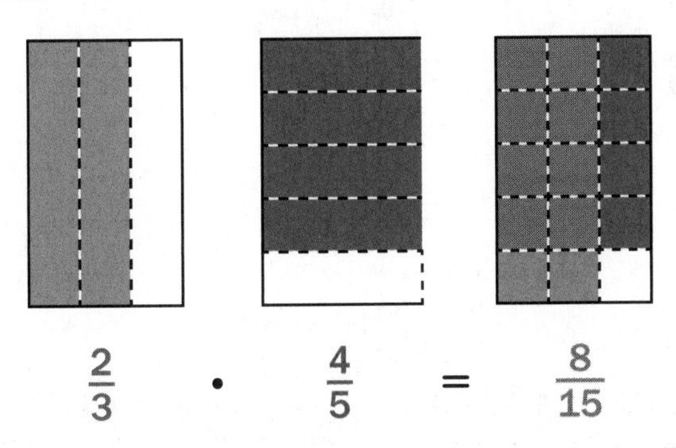

$$\frac{2}{3} \cdot \frac{4}{5} = \frac{8}{15}$$

The numerator is $2 \cdot 4 = 8$. The denominator is $3 \cdot 5 = 15$.

VOCABULARY

multiplication
an operation on two numbers to find their product; repeated addition

When multiplying a fraction by a whole number, write the whole number with a denominator of 1. For example, $3 = \frac{3}{1}$.

Example 1

Use models to multiply $4 \cdot \frac{2}{3}$. Write the product in simplest form.

1. Model $\frac{2}{3}$.
2. Repeat 4 times.
3. There are 8 shaded thirds, or $\frac{8}{3}$.

$$\frac{8}{3} = 2\frac{2}{3}$$

YOUR TURN!

Use models to multiply $3 \cdot \frac{3}{4}$. Write the product in simplest form.

1. Model ________.
2. Repeat ________ times.
3. There are ________ shaded ________, or ________.

________ = ________

GO ON

Example 2

Use models to multiply $\frac{1}{2} \cdot \frac{2}{3}$. Write the product in simplest form.

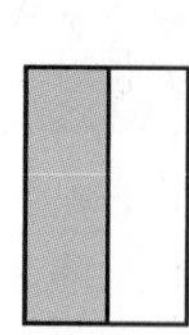

1. Model $\frac{1}{2}$.
2. Model $\frac{2}{3}$ on the same rectangle.
3. Count the total number of parts. **6**
4. Count the parts with both patterns. **2**

$$\frac{1}{2} \cdot \frac{2}{3} = \frac{2}{6} \text{ or } \frac{1}{3}$$

YOUR TURN!

Use models to multiply $\frac{3}{4} \cdot \frac{4}{5}$. Write the product in simplest form.

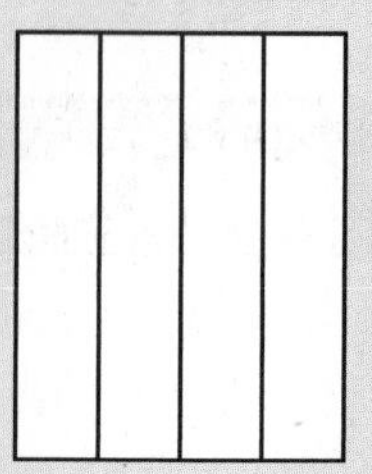

1. Model ______.
2. Model ______ on the same rectangle.

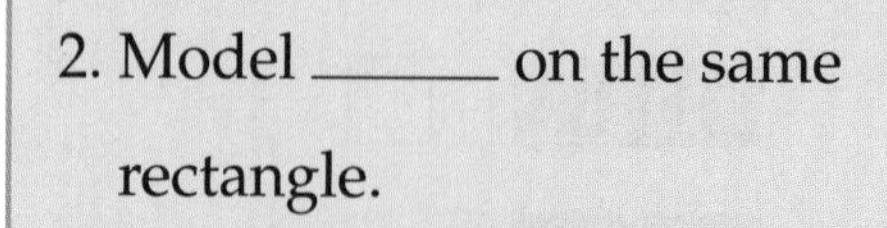

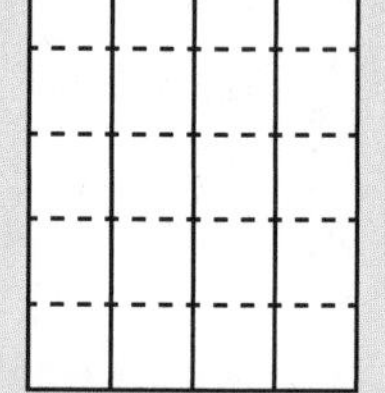

3. Count the total number of parts. ______
4. Count the parts with both patterns. ______

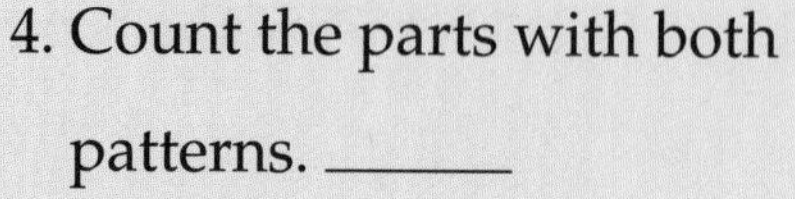

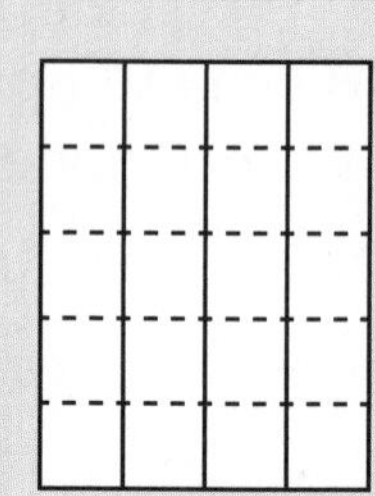

$\frac{3}{4} \cdot \frac{4}{5} =$ ______ or ______

Example 3

Multiply $\frac{3}{8} \cdot \frac{7}{5}$. Write the product in simplest form.

1. Multiply the numerators. Multiply the denominators.

$$\frac{3}{8} \cdot \frac{7}{5} = \frac{21}{40}$$

2. Is the fraction in simplest form? **Yes, there are no common factors. The product is $\frac{21}{40}$.**

YOUR TURN!

Multiply $\frac{5}{6} \cdot \frac{8}{11}$. Write the product in simplest form.

1. Multiply the numerators. Multiply the denominators.

$\frac{5}{6} \cdot \frac{8}{11} =$ ______

2. Is the fraction in simplest form? ______

____

Guided Practice

Use models to multiply.

1 $3 \cdot \frac{4}{5} =$ ______

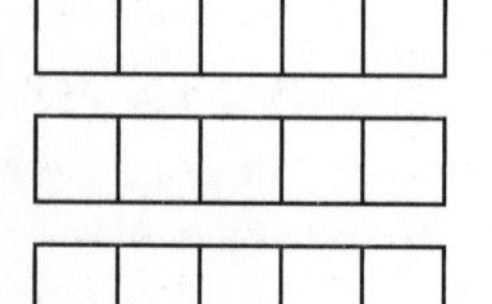

2 $4 \cdot \frac{1}{3} =$ ______

3 $\frac{1}{2} \cdot \frac{1}{4} =$ ______

4 $\frac{2}{3} \cdot \frac{3}{4} =$ ______

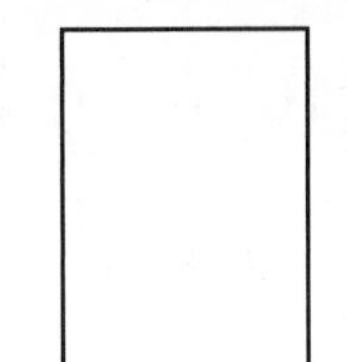

Step by Step Practice

5 Multiply $\frac{4}{5} \cdot \frac{5}{8}$. Write the product in simplest form.

Step 1 Multiply the numerators. Multiply the denominators.

$\frac{4}{5} \cdot \frac{5}{8} =$ ______

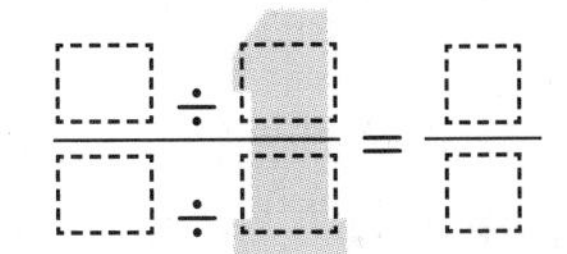

Step 2 Is the fraction in simplest form? ______.

Multiply. Write each product in simplest form.

6 $\frac{1}{5} \cdot \frac{2}{7} =$ ______

7 $\frac{11}{16} \cdot \frac{4}{7} =$ ______ $=$ ______

Step by Step Problem-Solving Practice

Solve.

8 **THEATER** The students in the ninth grade at Wilson High School who sold 10 tickets for the Spring play earned a spirit day. The ninth grade class has 128 students and $\frac{5}{8}$ of those students earned a spirit day. How many students earned a spirit day?

$128 \cdot \frac{5}{8} = \frac{\square}{\square}$

$= \frac{\square \div \square}{\square \div \square} = \frac{\square}{\square} =$ ______

Check off each step.

______ **Understand: I underlined key words.**

______ **Plan: To solve the problem, I will** ______.

______ **Solve: The answer is** ______.

______ **Check: I checked my answer by** ______.

GO ON

Skills, Concepts, and Problem Solving

Multiply. Write each product in simplest form.

9. $18 \cdot \frac{1}{2} =$ ______

10. $\frac{4}{5} \cdot \frac{3}{5} =$ ______

11. $\frac{7}{10} \cdot \frac{2}{3} =$ ______

12. $\frac{1}{4} \cdot 2\frac{3}{7} =$ ______

13. $\frac{2}{5} \cdot \frac{15}{16} =$ ______

14. $12 \cdot \frac{2}{5} =$ ______

15. $3\frac{1}{3} \cdot 4\frac{5}{6} =$ ______

16. $23 \cdot \frac{1}{6} =$ ______

17. $\frac{3}{8} \cdot \frac{8}{11} =$ ______

Solve. Write the answer in simplest form.

18. **SHOPPING** Mindy wants to buy a shirt that is on sale for $\frac{3}{4}$ of its original price of $36.00. What is the sale price of the shirt?

____

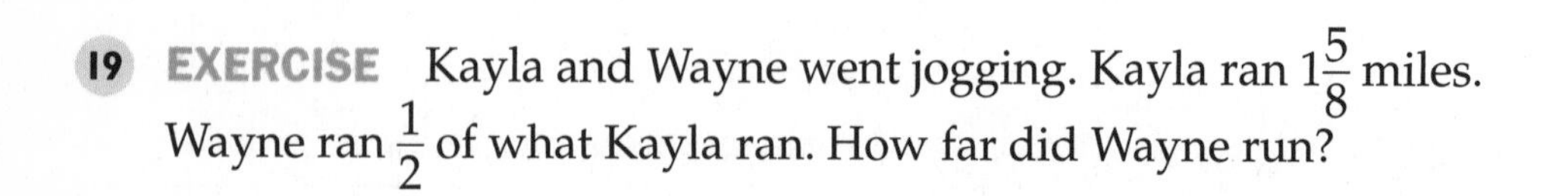

19. **EXERCISE** Kayla and Wayne went jogging. Kayla ran $1\frac{5}{8}$ miles. Wayne ran $\frac{1}{2}$ of what Kayla ran. How far did Wayne run?

____

20. **ENROLLMENT** Wilson High School has 568 students enrolled this year. Of those students, $\frac{3}{8}$ of the them are boys. How many students are boys?

____

Vocabulary Check **Write the vocabulary word that completes each sentence.**

21. ______________________ is an operation on two numbers to find their product.

22. **Reflect** When two mixed numbers are multiplied, is their product *always, sometimes* or *never* greater than 1? Explain.

____

____

____

STOP

Lesson 3-6

Divide Fractions

KEY Concept

A whole number can be divided by a fraction. You can use a model to show this. For example, to divide 3 by $\frac{1}{3}$, draw 3 rectangles.

1	2	3
4	5	6
7	8	9

Draw 3 rectangles. Divide each in thirds. Count the number of thirds in the 3 wholes.

So, $3 \div \frac{1}{3} = 9$.

To divide a fraction by another fraction, multiply by the **reciprocal**.

$$\frac{4}{9} \div \frac{3}{7} = \frac{4}{9} \cdot \frac{7}{3} = \frac{28}{27}$$

The reciprocal of $\frac{3}{7}$ is $\frac{7}{3}$.

VOCABULARY

division
an operation on two numbers where the first number is separated into the same number of equal groups as the second number

reciprocals
any two fractions that have a product of 1

Another name for a reciprocal is a multiplicative inverse.

Example 1

Use models to divide $2 \div \frac{1}{6}$. Write the quotient in simplest form.

1. Model 2 wholes.
2. Divide each whole into 6 equal parts.
3. Count the parts.

$2 \div \frac{1}{6} = 12$

YOUR TURN!

Use models to divide $3 \div \frac{1}{2}$. Write the quotient in simplest form.

1. Model ______ wholes.
2. Divide each whole into ______ equal parts.
3. Count the parts.

$3 \div \frac{1}{2} =$ ______

GO ON

Example 2

Divide $\frac{4}{7} \div \frac{5}{8}$. Write the quotient in simplest form.

1. Multiply $\frac{4}{7}$ by the reciprocal of $\frac{5}{8}$.

2. The reciprocal of $\frac{5}{8}$ is $\frac{8}{5}$.

$$\frac{4}{7} \cdot \frac{8}{5} = \frac{32}{35}$$

YOUR TURN!

Divide $\frac{3}{5} \div \frac{9}{13}$. Write the quotient in simplest form.

1. Multiply _____ by the reciprocal of _____.

2. The reciprocal of $\frac{9}{13}$ is _____.

_____ • _____ = __________

Example 3

Divide $\frac{3}{4} \div 2$. Write the quotient in simplest form.

1. Rewrite with all fractions.

$$\frac{3}{4} \div \frac{2}{1}$$

Convert 2 to $\frac{2}{1}$.

2. Multiply $\frac{3}{4}$ by the reciprocal of $\frac{2}{1}$.

The reciprocal of $\frac{2}{1}$ is $\frac{1}{2}$.

$$\frac{3}{4} \cdot \frac{1}{2} = \frac{3}{8}$$

YOUR TURN!

Divide $\frac{2}{3} \div 4$. Write the quotient in simplest form.

1. Rewrite with all fractions.

____

2. Multiply _____ by the reciprocal of _____.

The reciprocal of $\frac{4}{1}$ is _____.

_____ • _____ = _____

Guided Practice

Use models to divide. Write the quotient in simplest form.

1. $4 \div \frac{1}{3} =$ _____

2. $2 \div \frac{1}{5} =$ _____

Step by Step Practice

3 Divide $\frac{2}{7} \div \frac{3}{5}$. Write the quotient in simplest form.

Step 1 Multiply ______ by the reciprocal of ______.

The reciprocal is ______.

Step 2 ______ $\cdot$ ______ $=$ ______

Divide. Write each quotient in simplest form.

4 $\frac{5}{8} \div \frac{5}{16}$

______ $\cdot$ ______ $=$ ______________

5 $\frac{1}{4} \div \frac{7}{8}$

______ $\cdot$ ______ $=$ ______________

6 $\frac{3}{8} \div \frac{5}{9}$

______ $\cdot$ ______ $=$ ______

7 $\frac{2}{5} \div \frac{6}{10}$

______ $\cdot$ ______ $=$ ______________

Step by Step Problem-Solving Practice

Solve.

8 **SNACKS** Teresa has 6 cups of trail mix to divide into $\frac{1}{3}$ cup servings. How many portions will she make?

$6 \div \frac{1}{3} =$ ______ $\cdot$ ______ $=$ __________

Check off each step.

______ **Understand: I underlined key words.**

______ **Plan: To solve the problem, I will** ________________________________.

______ **Solve: The answer is** ________________________________.

______ **Check: I checked my answer by** ________________________________.

GO ON

Skills, Concepts, and Problem Solving

Divide. Write each quotient in simplest form.

9. $\frac{5}{12} \div \frac{3}{5} =$ ______

10. $\frac{5}{8} \div \frac{4}{5} =$ ______

11. $\frac{5}{16} \div \frac{5}{8} =$ ______

12. $\frac{14}{15} \div \frac{4}{5} =$ ______

13. $\frac{7}{9} \div \frac{2}{3} =$ ______

14. $2\frac{11}{20} \div 3\frac{2}{5} =$ ______

15. $3\frac{1}{5} \div 1\frac{7}{15} =$ ______

16. $2\frac{1}{2} \div 2\frac{2}{9} =$ ______

17. $7\frac{1}{3} \div 1\frac{5}{6} =$ ______

Solve. Write the answer in simplest form.

18. **PACKAGING** A box of individual pretzel bags contains 50 ounces. If the box holds 8 bags, how much does each bag hold?

____

19. **PARTY** At Natasha's birthday party, guests ate $6\frac{3}{4}$ pounds of watermelon. If the watermelon was divided evenly between the 9 girls at the party, how much watermelon did each girl eat?

____

20. **CROSS COUNTRY** Girls from the cross country team are running a $9\frac{1}{2}$ mile relay. If four girls participate in the relay, how far will each girl run?

____

Vocabulary Check **Write the vocabulary word that completes each sentence.**

21. ______________ is an operation on two numbers where the first number is separated into the same number of equal groups as the second number.

22. A fraction made from another fraction by switching the numerator and denominator is called a ______________.

23. **Reflect** When two fractions less than 1 are divided, is their quotient *always, sometimes* or *never* greater than 1? Explain.

____

____

STOP

Chapter 3

Progress Check 3 (Lessons 3-5 and 3-6)

Multiply. Write each product in simplest form.

1. $16 \cdot \frac{1}{4} =$ ______
2. $\frac{6}{9} \cdot \frac{8}{10} =$ ______
3. $\frac{5}{6} \cdot \frac{6}{7} =$ ______
4. $\frac{5}{7} \cdot 2\frac{3}{7} =$ ______
5. $\frac{5}{6} \cdot \frac{5}{6} =$ ______
6. $5 \cdot \frac{4}{5} =$ ______
7. $5\frac{1}{3} \cdot 3\frac{3}{5} =$ ______
8. $20 \cdot \frac{5}{12} =$ ______
9. $\frac{3}{10} \cdot \frac{5}{7} =$ ______

Divide. Write each quotient in simplest form.

10. $\frac{7}{16} \div \frac{3}{4} =$ ______
11. $\frac{6}{7} \div \frac{2}{3} =$ ______
12. $\frac{21}{25} \div \frac{3}{5} =$ ______
13. $\frac{9}{12} \div \frac{1}{3} =$ ______
14. $\frac{5}{18} \div \frac{7}{9} =$ ______
15. $4\frac{1}{10} \div 2\frac{3}{5} =$ ______
16. $2\frac{3}{4} \div 1\frac{3}{20} =$ ______
17. $5\frac{4}{5} \div 3\frac{5}{15} =$ ______
18. $2\frac{5}{9} \div 1\frac{5}{6} =$ ______

Solve. Write the answer in simplest form.

19. **READING** A book has 336 pages. If Shawn reads 22 pages each day, how many days will it be before he has finished the book?

____

20. **MEASUREMENT** Miguel uses $2\frac{3}{4}$ yards of canvas to make curtains. He uses $\frac{1}{2}$ as much to make a flag. How many yards of canvas does he use to make the flag?

____

21. **MOVIES** Zoe empties $3\frac{5}{9}$ boxes of popcorn into bowls. If each bowl holds $\frac{1}{3}$ a box of popcorn, how many bowls will she fill?

____

Lesson 3-7 Least Common Multiples

KEY Concept

A **multiple** is the product of a number and any whole number.

Multiples of 5:

$5 \cdot 1 = 5$	$5 \cdot 4 = 20$	$5 \cdot 7 = 35$
$5 \cdot 2 = 10$	$5 \cdot 5 = 25$	$5 \cdot 8 = 40$
$5 \cdot 3 = 15$	$5 \cdot 6 = 30$	$5 \cdot 9 = 45$

To find the **least common multiple**, or LCM, you can list the multiples of each number.

multiples of 12: 12, 24, 36, 48 . . .
multiples of 9: 9, 18, 27, 36 . . .
LCM of 9 and 12: 36

You can also use factor trees to find the LCM. First, use a factor tree to find the **prime factorization**.

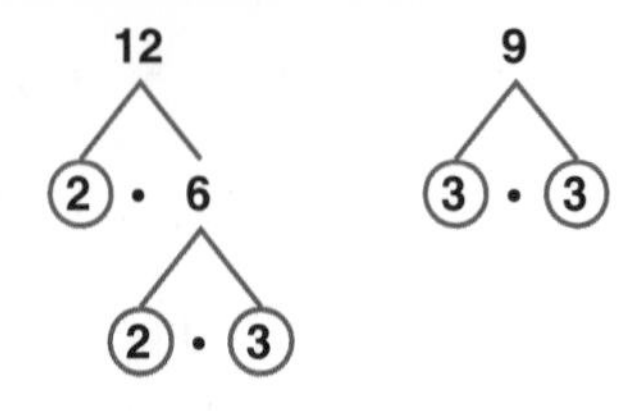

The prime factorization of 12 is $2 \cdot 2 \cdot 3$.
The prime factorization of 9 is $3 \cdot 3$.

The LCM is a product of all the factors of both numbers. Use the common prime factors only once.

So, the LCM of 12 and 9 is $2 \cdot 2 \cdot 3 \cdot 3 = 36$.

VOCABULARY

least common multiple (LCM)
the least whole number greater than 1 that is a common multiple of each of two more numbers

multiple
the product of a number and any whole number

prime factorization
a whole number as product of its prime factors

Find the LCM of more than two numbers using either method.

Example 1

Find the LCM of 12 and 20 by listing the multiples.

1. List the multiples of each number.

   12: 12, 24, 36, 48, 60, 72, 84, 96

   20: 20, 40, 60, 80

2. The LCM is 60.

YOUR TURN!

Find the LCM of 7 and 13 by listing the multiples.

1. List the multiples of each number.

   7: ______________________________

   ____

   13: ______________________________

2. The LCM is ______.

Example 2

Find the LCM of 28 and 32 by using prime factorization.

1. Write the prime factorization of each number using factor trees.

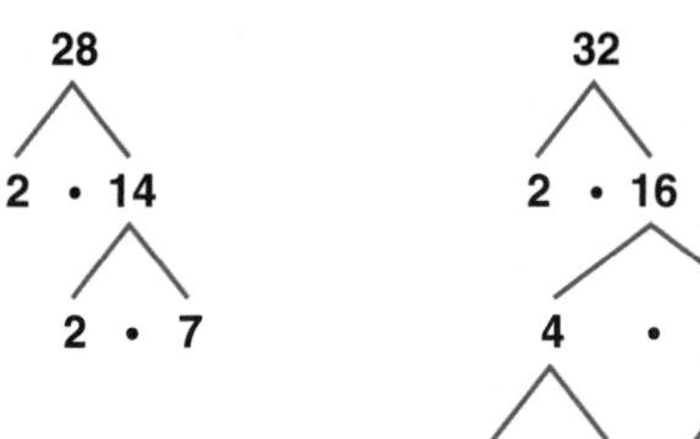

2. List the prime factors from the factor trees. Circle the common prime factors.

   28 = 2 • 2 • 7

   32 = 2 • 2 • 2 • 2 • 2

3. Multiply each factor. Use common factors only once.

   2 • 2 • 7 • 2 • 2 • 2 = 224

4. The LCM is 224.

YOUR TURN!

Find the LCM of 24 and 25 by using prime factorization.

1. Write the prime factorization of each number using factor trees.

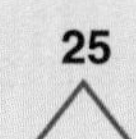

2. List the prime factors from the factor trees. Circle the common prime factors.

   24 = ______________________

   25 = ______________________

3. Multiply each factor. Use common factors only once.

   ____

4. The LCM is ______.

Find the LCM of each pair.

1. 18 and 21

   multiples of 18: ______________________

   multiples of 21: ______________________

   LCM: ______

2. 24 and 36

   multiples of 24: ______________________

   multiples of 36: ______________________

   LCM: ______

GO ON

Step by Step Practice

3 Find the LCM of 5 and 9 by listing mutliples.

Step 1 List the multiples of each number.

5: ______

9: ______

Step 2 Circle the smallest multiple the lists have in common.

Step 3 The LCM is ____.

Guided Practice

Find the LCM of each pair by using prime factorization.

4 16 and 20 ____

prime factors of 16: ______

prime factors of 20: ______

LCM: ______

16 20

5 24 and 32 ____

prime factors of 24: ______

prime factors of 32: ______

LCM: ______

24 32

6 4 and 17

prime factors of 4: ______

prime factors of 17: ______

LCM: ______

4 17

Find the LCM of each pair.

7. 10 and 16 ______

8. 28 and 42 ______

9. 18 and 22 ______

10. 7 and 30 ______

Step by Step Problem-Solving Practice

Solve.

11. **PARTY FAVORS** Kevin is buying party favors for a party. Paddleballs are sold in bags of 10, chocolate coins are sold in packages of 8, and glow in the dark bracelets are sold in sets of 12. Kevin wants to have an equal number of paddleballs, chocolate coins, and glow bracelets. What is the least number of each item he needs to buy?

multiples of 8: ______________________________

multiples of 10: ______________________________

multiples of 12: ______________________________

Check off each step.

______ **Understand: I underlined key words.**

______ **Plan: To solve the problem, I will** ______________________________.

______ **Solve: The answer is** ______________________________.

______ **Check: I checked my answer by** ______________________________.

prime factors of 8: ______________________________

prime factors of 12: ______________________________

prime factors of 10: ______________________________

____

GO ON

Skills, Concepts, and Problem Solving

Find the LCM for each pair.

12 6 and 8 ______ 13 11 and 12 ______ 14 6 and 7 ______

15 14 and 21 ______ 16 22 and 33 ______ 17 16 and 36 ______

18 24 and 64 ______ 19 18 and 90 ______ 20 90 and 63 ______

Solve.

SNACKS Ms. Kline needs to buy snacks and drinks for Jaime's preschool class. Fruit snacks come in boxes of 6. Juice boxes come in packages of 8. Use this information to answer Exercises 21 and 22.

21 If Ms. Kline wants to have the same number of juice boxes and fruit snacks, what is the least number of items that she should buy?

____

22 A box of fruit snacks costs $3.15 and a package of juice boxes costs $5.50. How much will Ms. Kline pay for the snacks and drinks?

____

Vocabulary Check **Write the vocabulary word that completes each sentence.**

23 The product of a number and any whole number is a(n)

______________________ of that number.

24 Expressing a composite number as a product of its prime factors is

called ______________________.

25 The ______________________ is the least whole number greater than 1 that is a common multiple of each of two more numbers.

26 **Reflect** Dana says that the easiest way to find the LCM of two numbers is to multiply them together. For example, the LCM of 6 and 5 is found by $6 \cdot 5 = 30$. 30 is a multiple of both 5 and 6. Does Dana's method always work when finding the LCM? Explain.

____

____

____

STOP

Lesson 3-8

Add Fractions

KEY Concept

To add fractions with **like denominators**, add the numerators.

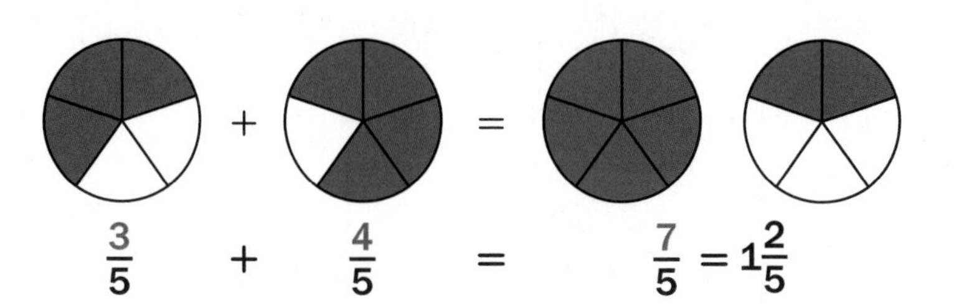

To add fractions with **unlike denominators**, rewrite the fractions so their denominators are the same. Use the least common multiple to find the **least common denominator**.

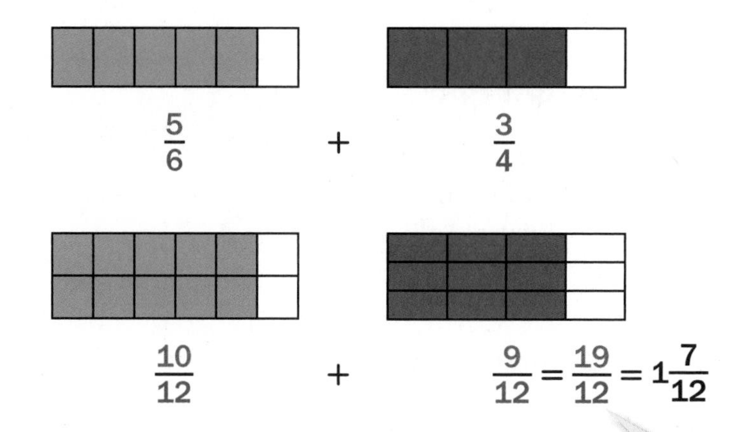

The LCD of 6 and 4 is 12.

VOCABULARY

like denominators
denominators that are the same

least common denominators (LCD)
the least common multiple of the denominators of two or more fractions

unlike denominators
denominators that are not the same

To add mixed numbers, convert each mixed number to an improper fraction, then add the improper fractions. For all addition problems, be sure the answers are written in simplest form.

GO ON

Example 1

Add $\frac{1}{3} + \frac{2}{3}$.

1. Model each fraction.

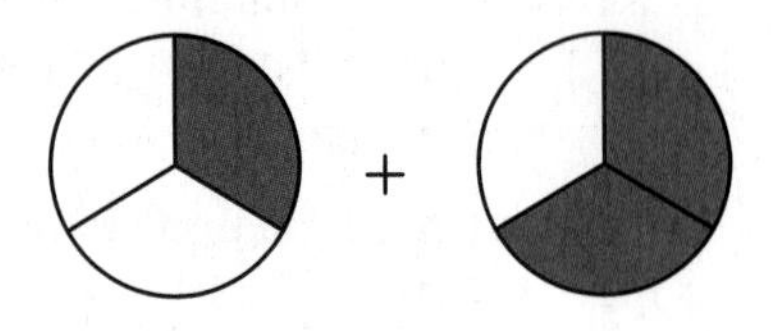

2. Count the total number of thirds.

$\frac{3}{3}$

3. Write your answer in simplest form.

$\frac{3}{3} = 1$

YOUR TURN!

Add $\frac{3}{4} + \frac{2}{4}$.

1. Model each fraction.

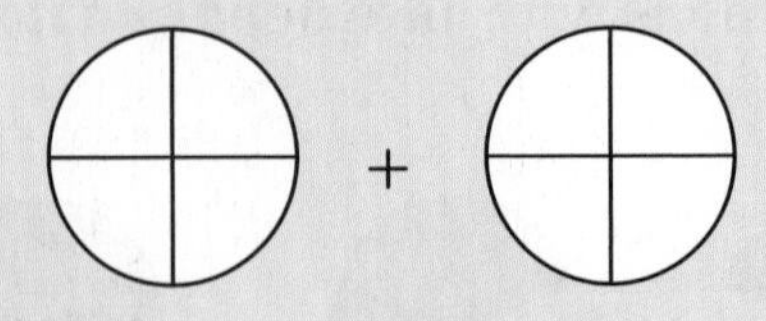

2. Count the total number of ________.

____

3. Write your answer in simplest form.

________ = ________

Example 2

Add $\frac{7}{11} + \frac{16}{11}$.

1. Are the denominators the same?

yes

2. Add the numerators.

$7 + 16 = 23$

3. Write the fraction.

$\frac{23}{11}$

4. Write your answer in simplest form.

$\frac{23}{11} = 2\frac{1}{11}$

YOUR TURN!

Add $\frac{7}{10} + \frac{19}{10}$.

1. Are the denominators the same?

____

2. Add the numerators.

____

3. Write the fraction.

____

4. Write your answer in simplest form.

________ = ________

Example 3

Add $1\frac{1}{6} + 2\frac{7}{15}$.

1. Convert the mixed numbers to improper fractions.

   $1\frac{1}{6} = \frac{7}{6}$ $\quad 2\frac{7}{15} = \frac{37}{15}$

2. Are the denominators the same?

   no

3. List the multiples of 6 and 15 to find the LCD.

   multiples of 6: 6, 12, 18, 24, 30
   multiples of 15: 15, 30
   LCD: 30

4. Rewrite the fractions with 30 in the denominators.

   $\frac{7 \cdot 5}{6 \cdot 5} = \frac{35}{30}$ $\quad \frac{37 \cdot 2}{15 \cdot 2} = \frac{74}{30}$

5. Add the fractions. Convert the sum to a mixed number.

   $\frac{35}{30} + \frac{74}{30} = \frac{109}{30} = 3\frac{19}{30}$

YOUR TURN!

Add $3\frac{3}{4} + 2\frac{5}{6}$.

1. Convert the mixed numbers to improper fractions.

   $3\frac{3}{4} =$ ________ $\quad 2\frac{5}{6} =$ ________

2. Are the denominators the same?

   ____

3. List the multiples of ________ and ________ to find the LCD.

   multiples of 4: ________________
   multiples of 6: ________________
   LCD: ________________

4. Rewrite the fractions with ______ in the denominators.

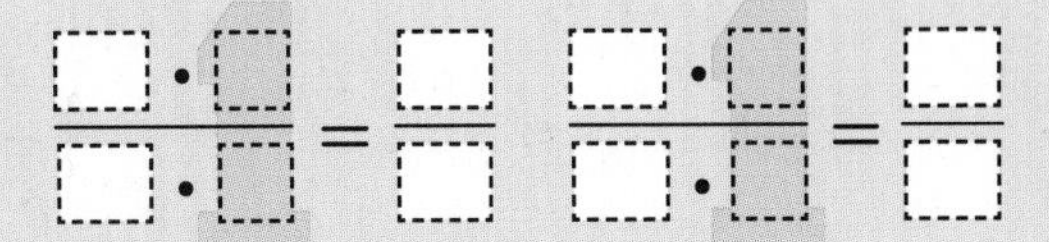

5. Add the fractions. Convert the sum to a mixed number.

   ____

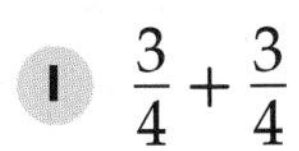

Guided Practice

Add.

1 $\frac{3}{4} + \frac{3}{4}$

Model each fraction.

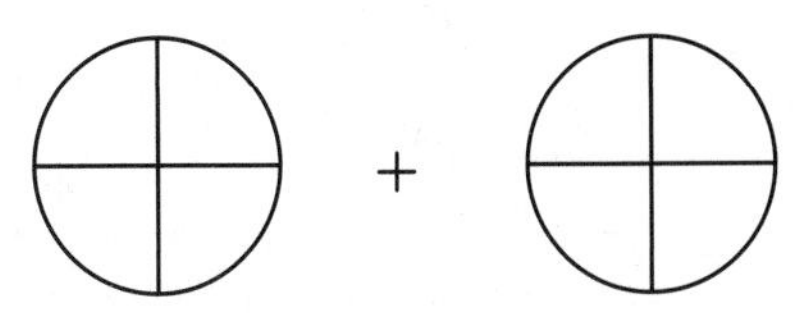

Count the total number of ________.

________ = ________

2 $\frac{1}{2} + \frac{1}{2}$

Model each fraction.

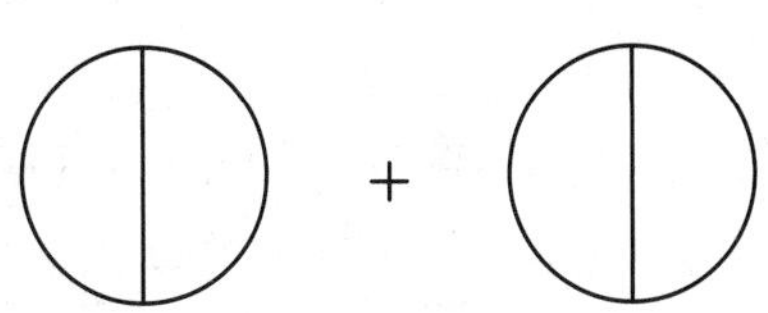

Count the total number of ________.

________ = ________

GO ON

Add.

3. $\frac{5}{10} + \frac{2}{10} =$ ______

4. $\frac{2}{6} + \frac{1}{6} =$ ______ $=$ ______

Step by Step Practice

5. **Add $3\frac{1}{5} + 2\frac{3}{4}$.**

Step 1 Convert the mixed numbers to improper fractions.

$3\frac{1}{5} =$ ______ $2\frac{3}{4} =$ ______

Step 2 List the multiples of each denominator.

multiples of 5: ______

multiples of 4: ______

Step 3 Find the LCD. ______

Step 4 Rewrite the fractions with ______ in the denominator.

$5 \cdot$ ______ $=$ ______ $\frac{\square \cdot \square}{5 \cdot \square} = \frac{\square}{\square}$

$4 \cdot$ ______ $=$ ______ $\frac{\square \cdot \square}{4 \cdot \square} = \frac{\square}{\square}$

Step 5 Add the fractions and convert to a mixed number.

______ $=$ ______

Add.

6. $2\frac{3}{8} + 4\frac{11}{20}$

LCD: ______

$\frac{19 \cdot \square}{8 \cdot \square} = \frac{\square}{\square}$ $\frac{91 \cdot \square}{20 \cdot \square} = \frac{\square}{\square}$

______ $+$ ______ $=$ ______ $=$ ______

7. $5\frac{5}{11} + 3\frac{3}{4}$

LCD: ______

$\frac{60 \cdot \square}{11 \cdot \square} = \frac{\square}{\square}$ $\frac{15 \cdot \square}{4 \cdot \square} = \frac{\square}{\square}$

______ $+$ ______ $=$ ______ $=$ ______

Step by Step Problem-Solving Practice

Solve.

8 **BIKING** Lynn and Daniela biked two different trails at National Falls Park. The first trail was $4\frac{5}{8}$ miles and the second trail was $3\frac{1}{6}$ miles. How far did they bike on both trails combined?

$4\frac{5}{8} + 3\frac{1}{6} =$ ____ + ____ = ____ + ____ = ____ = ____

Check off each step.

____ Understand: I underlined key words.

____ Plan: To solve the problem, I will ____________.

____ Solve: The answer is ____________.

____ Check: I checked my answer by ____________.

Skills, Concepts, and Problem Solving

Add.

9 $\frac{1}{7} + \frac{4}{7} =$ ____

10 $\frac{5}{12} + \frac{3}{12} =$ ____

11 $\frac{7}{20} + \frac{9}{20} =$ ____

12 $\frac{3}{7} + \frac{4}{5} =$ ____

13 $\frac{1}{6} + \frac{2}{9} =$ ____

14 $\frac{3}{11} + \frac{1}{3} =$ ____

15 $5\frac{1}{2} + 3\frac{2}{3} =$ ____

16 $8\frac{1}{2} + 3\frac{4}{5} =$ ____

17 $2\frac{5}{12} + 4\frac{11}{15} =$ ____

18 $1\frac{2}{5} + 7\frac{3}{10} =$ ____

19 $6\frac{11}{12} + 2\frac{1}{6} =$ ____

20 $9\frac{7}{8} + 3\frac{1}{8} =$ ____

GO ON

Solve. Write the answer in simplest form.

21 **CONDITIONING** Marcus is conditioning for soccer season. He runs $2\frac{1}{2}$ miles each morning and $3\frac{1}{4}$ each evening. How far does he run each day?

____

22 **FRUIT** Mrs. Wright bought fresh fruit at the local market. She bought $1\frac{1}{4}$ pounds of apples, $\frac{5}{8}$ pound of pears, and $1\frac{2}{3}$ pounds of peaches. How many pounds of fruit did she buy?

____

23 **SCHEDULE** Tony's schedule before school includes $\frac{1}{2}$ of an hour of exercise, $\frac{3}{4}$ of an hour to shower and get dressed, and $\frac{1}{5}$ of an hour to eat breakfast. How much time does Tony need in the morning to complete his schedule?

____

Vocabulary Check **Write the vocabulary word that completes each sentence.**

24 Denominators that are not the same are ______________________.

25 Denominators that are the same are ______________________.

26 ______________________ are the least common multiple of the denominators of two or more fractions.

27 **Reflect** Alana has found the sum of $5\frac{1}{2} + 3\frac{2}{3}$. She added the numerators together, the denominators together, and the whole numbers together. Her sum is $8\frac{3}{5}$. Is she correct? Explain.

____

____

____

STOP

Lesson 3-9

Subtract Fractions

KEY Concept

To subtract fractions with **like denominators**, subtract the numerators.

$$\frac{4}{5} - \frac{3}{5} = \frac{1}{5}$$

To subtract fractions with **unlike denominators**, rewrite the fractions so their denominators are the same. Use the least common multiple to find the **least common denominator**.

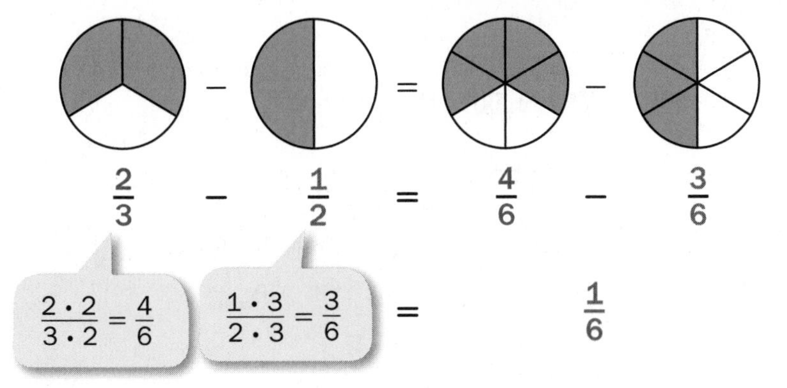

To subtract mixed numbers, you must have like denominators. Sometimes, you also have to rewrite the numerators.

$$7\frac{2}{3} \longrightarrow 7\frac{2}{3} = 7\frac{2 \cdot 8}{3 \cdot 8} \longrightarrow 7\frac{16}{24}$$

$$-5\frac{7}{8} \longrightarrow 5\frac{7}{8} = 5\frac{7 \cdot 3}{8 \cdot 3} \longrightarrow -5\frac{21}{24}$$

16 − 21 is a negative number. Rewrite $7\frac{16}{24}$ so that its numerator is greater than 21.

$$7\frac{16}{24} = 6 + 1 + \frac{16}{24} = 6 + \frac{24}{24} + \frac{16}{24} = 6\frac{40}{24}$$

Rewrite 1 as a fraction that you can add to $\frac{16}{24}$.

$$6\frac{40}{24} - 5\frac{21}{24} = 1\frac{19}{24}$$

VOCABULARY

least common denominators (LCD)
the least common multiple of the denominators of 2 or more fractions

like denominators
denominators that are the same

unlike denominators
denominators that are not the same

GO ON

Example 1

Subtract $\frac{5}{7} - \frac{3}{7}$ using models.

1. Model the first fraction. Then cross out the amount from the second.

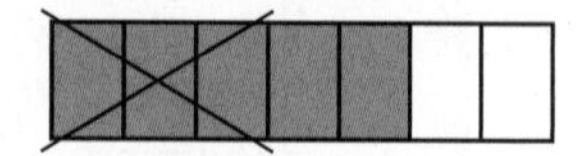

2. Count the total number of shaded sevenths that are left.

$\frac{2}{7}$

YOUR TURN!

Subtract $\frac{8}{9} - \frac{3}{9}$ using models.

1. Model the first fraction. Then cross out the amount from the second.

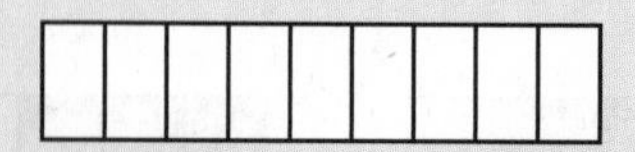

2. Count the total number of shaded

_______ that are left. _______

Example 2

Subtract $4\frac{7}{15} - 3\frac{8}{15}$.

1. Write one whole as a fraction using an equivalent form of one.

$1 = \frac{15}{15}$

2. Rewrite $4\frac{7}{15}$ using this number.

$4 + \frac{7}{15} = 3 + \frac{15}{15} + \frac{7}{15}$

$= 3\frac{22}{15}$

3. Subtract the whole numbers.

$3 - 3 = 0$

4. Subtract the fractions.

$\frac{22}{15} - \frac{8}{15} = \frac{14}{15}$

5. Add the differences.

$4\frac{7}{15} - 3\frac{8}{15} = 0 + \frac{14}{15}$

$= \frac{14}{15}$

YOUR TURN!

Subtract $2\frac{1}{4} - 1\frac{3}{4}$.

1. Write one whole as a fraction using an equivalent form of one.

1 = _______

2. Rewrite $2\frac{1}{4}$ using this number.

$2 + \frac{1}{4} = 1 +$ _______ + _______

= _______

3. Subtract the whole numbers.

_______ = _______

4. Subtract the fractions.

_______ = _______

5. Add the differences.

$2\frac{1}{4} - 1\frac{3}{4} =$ _______ + _______

= _______

Example 3

Subtract $5\frac{2}{5} - 2\frac{5}{12}$.

1. List the multiples of 5 and 12 to find the LCD.

   5: 5, 10, 15, 20, 25, 30, 35, 40, 45, 50, 55, 60
   12: 12, 24, 36, 48, 60

2. The LCD is 60. Rewrite the fractions with 60 in the denominators.

   $5\frac{24}{60} - 2\frac{25}{60}$

3. Rewrite $5\frac{24}{60}$ using an equivalent form of one.

   $5\frac{24}{60} = 4 + \frac{60}{60} + \frac{24}{60}$
   $= 4\frac{84}{60}$

4. Subtract.

   $4\frac{84}{60} - 2\frac{25}{60} = 2\frac{59}{60}$

YOUR TURN!

Subtract $6\frac{1}{3} - 3\frac{3}{7}$.

1. List the multiples of ______ and ______.

   3: ____________________

   7: ____________________

2. The LCD is ______. Rewrite the fractions with ______ in the denominators.

   ______ − ______

3. Rewrite ______ using an equivalent form of one.

   ______ = ______ + $\frac{\square}{\square}$ + ______

   = ______

4. Subtract.

   ______ − ______ = ______

Guided Practice

Subtract using models.

1. $\frac{4}{6} - \frac{2}{6}$

Count the total number of shaded

______ left.

______

2. $\frac{3}{5} - \frac{1}{5}$

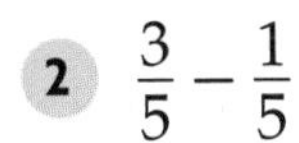

Count the total number of shaded

______ left.

______

GO ON

Subtract.

3 $3\frac{1}{5} - 1\frac{3}{5}$

$3 + \frac{1}{5} = 2 +$ ______ + ______ = ______

______ − ______ = ______

4 $6\frac{2}{9} - 2\frac{5}{9}$

$6 + \frac{2}{9} = 5 +$ ______ + ______ = ______

______ − ______ = ______

Step by Step Practice

5 Subtract $9\frac{4}{5} - 2\frac{1}{3}$.

Step 1 List the multiples of ______ and ______ to find the LCD.

5: ______________________________

3: ______________________________

Step 2 The LCD is ______. Rewrite the fractions with ______ in the denominators.

$9\frac{4}{5} - 2\frac{1}{3} =$ ______ − ______

Step 3 Subtract.

______ − ______ = ______

Subtract.

6 $4\frac{2}{3} - 3\frac{3}{16}$

3: ______________________________

____

16: ______________________________

LCD: ______________________________

Rewrite the fractions with ______ in the denominators and subtract.

______ − ______ = ______

7 $8\frac{3}{10} - 5\frac{12}{25}$

10: ______________________________

25: ______________________________

LCD: ______________________________

Rewrite the fractions with ______ in the denominators and subtract.

______ − ______ = ______

Step by Step Problem-Solving Practice

Solve.

8 **RACING** Nate entered a $13\frac{3}{5}$ mile dirt bike race. He completed $7\frac{3}{8}$ miles when his tire blew and he had to quit the race. How many miles were left to race when he quit?

$13\frac{3}{5} - 7\frac{3}{8} =$ ______ − ______ = ______

Check off each step.

______ **Understand: I underlined key words.**

______ **Plan: To solve the problem, I will** ______________________.

______ **Solve: The answer is** ______________________.

______ **Check: I checked my answer by** ______________________.

____

Skills, Concepts, and Problem Solving

Subtract.

9 $\frac{7}{8} - \frac{5}{8} =$ ______

10 $\frac{5}{6} - \frac{3}{6} =$ ______

11 $\frac{7}{12} - \frac{3}{12} =$ ______

12 $\frac{4}{5} - \frac{1}{3} =$ ______

13 $\frac{6}{7} - \frac{9}{14} =$ ______

14 $\frac{9}{10} - \frac{1}{3} =$ ______

15 $9\frac{5}{6} - 2\frac{2}{7} =$ ______

16 $6\frac{1}{2} - 2\frac{2}{15} =$ ______

17 $10\frac{7}{9} - 3\frac{1}{3} =$ ______

18 $5\frac{1}{9} - 3\frac{4}{5} =$ ______

19 $12\frac{2}{3} - 8\frac{8}{9} =$ ______

20 $7\frac{3}{5} - 2\frac{13}{20} =$ ______

GO ON

Solve. Write the answer in simplest form.

21 **DELI** Mr. Watson bought 4 pounds of deli ham for his family to make sandwiches. After the first day, his family had eaten $1\frac{3}{5}$ pounds of ham. How much ham was left for the rest of the week?

____

22 **HEIGHT** Carlos is $5\frac{2}{3}$ feet tall and his younger brother is $4\frac{5}{12}$ feet tall. How much taller is Carlos than his brother?

____

23 **CHARITY** Kerri wants to cut her hair and donate it for medical patients. Her hair is $18\frac{2}{3}$ inches long. She needs to cut off $10\frac{1}{2}$ inches in order to donate it. How long will her hair be after she cuts it?

____

Vocabulary Check **Write the vocabulary word that completes each sentence.**

24 The least common multiple of the denominators of two or more fractions is called the ______________________________.

25 ______________________________ are denominators that are the same.

26 ______________________________ are denominators that are not the same.

27 **Reflect** Linda found the difference of $4\frac{5}{9}$ and $2\frac{1}{3}$ to be $2\frac{4}{6}$. Explain what she did incorrectly.

____

____

____

____

STOP

Chapter 3

Progress Check 4 (Lessons 3-7, 3-8, and 3-9)

Find the LCM.

1 5 and 6 ______

2 6 and 9 ______

3 3 and 10 ______

4 12 and 18 ______

5 21 and 24 ______

6 14 and 15 ______

Add or subtract. Write the answer in simplest form.

7 $\frac{2}{5} + \frac{4}{5} =$ ______

8 $\frac{9}{10} - \frac{7}{10} =$ ______

9 $\frac{3}{8} - \frac{1}{8} =$ ______

10 $\frac{5}{6} + \frac{3}{4} =$ ______

11 $\frac{3}{5} + \frac{7}{10} =$ ______

12 $\frac{2}{3} + \frac{1}{4} =$ ______

13 $4\frac{3}{4} - 1\frac{1}{2} =$ ______

14 $5\frac{6}{7} + 3\frac{5}{14} =$ ______

15 $1\frac{5}{16} + 6\frac{5}{12} =$ ______

16 $2\frac{4}{13} - \frac{4}{13} =$ ______

17 $\frac{4}{7} + \frac{5}{7} =$ ______

18 $\frac{13}{18} - \frac{7}{18} =$ ______

Solve. Write the answer in simplest form.

19 **VETERINARIAN** A veterinarian weighed three puppies. One puppy weighed $3\frac{1}{6}$ pounds, another weighed $2\frac{7}{8}$ pounds, and the last weighed $3\frac{2}{3}$ pounds. How many pounds did the puppies weigh in all?

____

20 **LUNCH** During the first lunch period, students ate $11\frac{1}{3}$ pans of pizza. If there were $26\frac{4}{7}$ pizzas to begin with, how much pizza is left?

____

21 **FOOD** Jesse collects $35\frac{3}{4}$ ounces of maple syrup from his uncle's maple tree. The next day he collects $26\frac{7}{10}$ ounces. How much more did he collect the first day than the second?

____

Chapter 3

Chapter 3 Test

Find the GCF of each set of numbers.

1. 12, 30 ____________
2. 15, 45 ____________
3. 24, 36 ____________
4. 81, 63 ____________
5. 27, 15 ____________
6. 20, 25 ____________

Write each fraction in simplest form.

7. $\frac{4}{10} =$ ____________
8. $\frac{55}{100} =$ ____________
9. $\frac{9}{21} =$ ____________
10. $\frac{18}{24} =$ ____________
11. $\frac{8}{30} =$ ____________
12. $\frac{54}{63} =$ ____________

Write each improper fraction as a mixed number in simplest terms.

13. $\frac{8}{5} =$ ____________
14. $\frac{74}{9} =$ ____________
15. $\frac{65}{3} =$ ____________
16. $\frac{25}{6} =$ ____________
17. $\frac{34}{4} =$ ____________
18. $\frac{43}{7} =$ ____________

Write >, <, or = to compare each fraction pair.

19. $\frac{1}{9} \bigcirc \frac{1}{7}$
20. $\frac{8}{20} \bigcirc \frac{2}{5}$
21. $\frac{4}{9} \bigcirc \frac{4}{10}$

Order the fractions and mixed numbers from least to greatest.

22. $1\frac{1}{7}, \frac{7}{9}, \frac{13}{10}$ ____________________
23. $2\frac{1}{4}, \frac{7}{2}, 2\frac{3}{10}$ ____________________

Multiply. Write each product in simplest form.

24 $15 \cdot \frac{1}{3} =$ ____________

25 $\frac{2}{3} \cdot \frac{3}{8} =$ ____________

26 $\frac{5}{18} \cdot \frac{6}{11} =$ ____________

27 $\frac{2}{5} \cdot 3\frac{1}{3} =$ ____________

28 $\frac{4}{9} \cdot \frac{4}{9} =$ ____________

29 $10 \cdot \frac{4}{5} =$ ____________

Divide. Write each quotient in simplest form.

30 $\frac{3}{7} \div \frac{3}{5} =$ ____________

31 $\frac{1}{4} \div 2\frac{1}{2} =$ ____________

32 $\frac{63}{72} \div \frac{7}{9} =$ ____________

33 $\frac{5}{6} \div \frac{1}{3} =$ ____________

34 $\frac{7}{10} \div \frac{3}{5} =$ ____________

35 $6\frac{1}{3} \div 2\frac{1}{9} =$ ____________

Solve. Write the answer in simplest form.

36 **LONG JUMP** Lora jumped $4\frac{1}{6}$ meters. Write the distance Lora jumped as an improper fraction.

____

37 **BASEBALL** Evan got a hit in 52 at-bats. He was up to bat 180 times. What part of the times he batted did Evan get a hit?

____

Correct the mistake.

38 Tabitha did this division problem: $\frac{7}{12} \div \frac{6}{21}$. Her quotient was $\frac{1}{6}$. Determine what she did incorrectly. Correct the mistake to find the quotient.

____

____

____

STOP

Chapter 4

Real Numbers

What determines a good tennis player?

In tennis, a higher first serve percentage can help win more matches. 60% to 70% of successful serves during a match may indicate a good tennis player.

STEP 1 Chapter Pretest

Are you ready for Chapter 4?
Take the Chapter 4 Pretest to find out.

STEP 2 Preview

Get ready for Chapter 4. Review these skills and compare them with what you will learn in this chapter.

What You Know

You know how to write decimals in standard form.

Example: eight tenths = 0.8

TRY IT!

Write each decimal in standard form.

1. six tenths = ______
2. three and four tenths = ______
3. forty-five hundredths = ______

What You Will Learn

Lesson 4-2

You can write a decimal as a fraction by using the place value as the denominator.

100	10	1	0.1	0.01	0.001
hundreds	tens	ones	tenths	hundredths	thousandths
0	0	0.	9	0	0

Example: $0.9 = \frac{9}{10}$

What You Know

You know how to multiply a number by itself.

Example: $2 \cdot 2 = 4$

TRY IT!

Multiply.

4. $3 \cdot 3 =$ ______
5. $4 \cdot 4 =$ ______
6. $5 \cdot 5 =$ ______
7. $6 \cdot 6 =$ ______

What You Will Learn

Lesson 4-5

A **perfect square** is the product of a number multiplied by itself.

The **square root** is the number that was multiplied by itself.

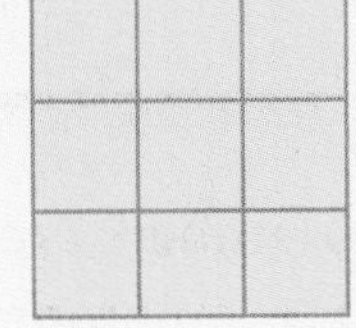

There are 3 tiles on each side of the square.
So, $\sqrt{9} = 3$.

What You Know

You know how to graph numbers.

Example: Graph 1 and 3.

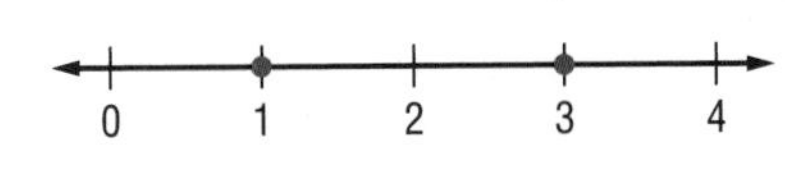

What You Will Learn

Lesson 4-6

To compare real numbers, use a number line.

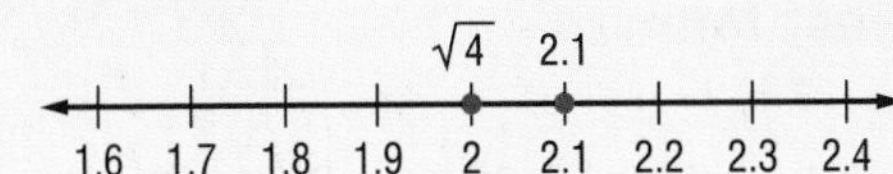

So, $2.1 > \sqrt{4}$.

Lesson 4-1

Rational Numbers

KEY Concept

A **rational number** is a number that can be written as a fraction.

-14 is a rational number because it can be written as $-\frac{14}{1}$.

$\frac{7}{0}$ is not a rational number because $7 \div 0$ is not defined.

Counting numbers, **whole numbers**, and **integers** are all rational numbers because they can be written as fractions.

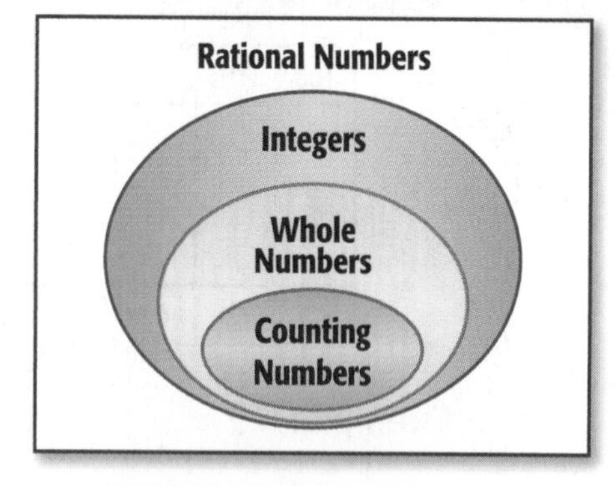

Terminating and **repeating decimals** are also rational numbers because they can be written as fractions.

Terminating decimals have an end.

The overbar means that the 3 repeats.

$0.25 = \frac{25}{100}$ $\quad 0.\overline{3} = \frac{1}{3}$

VOCABULARY

counting numbers
numbers used to count objects; also called natural numbers

integers
the set of whole numbers and their opposites

natural numbers
another name for counting numbers

rational number
the set of numbers expressed in the form of a fraction $\frac{a}{b}$, where a and b are integers and $b \neq 0$

repeating decimal
a decimal whose digits repeat in groups of one or more

terminating decimal
a decimal whose digits end; every terminating decimal can be written as a fraction with a denominator of 10, 100, 1,000, and so on

All numbers are rational numbers except non-repeating and non-terminating decimals.

Example 1

Write the number 13 as a fraction in simplest form.

1. Write 13 as a fraction with 1 as the denominator.

$13 = \frac{13}{1}$

YOUR TURN!

Write the number −7 as a fraction in simplest form.

1. Write ______ as a fraction with 1 as the denominator.

$-7 = \frac{\square}{\square} =$ ______

Example 2

Write the decimal −0.42 as a fraction in simplest form.

1. Write −0.42 as a fraction with 100 as the denominator.

$$-\frac{42}{100}$$

2. Simplify.

Divide by the GCF.

$$-\frac{42 \div 2}{100 \div 2} = -\frac{21}{50}$$

YOUR TURN!

Write the decimal 0.75 as a fraction in simplest form.

1. Write 0.75 as a as a fraction with 100 as the denominator.

$$\frac{\square}{100}$$

2. Simplify.

$$\frac{\square \div \square}{100 \div \square} = \frac{\square}{\square}$$

Example 3

Identify all sets to which the number −4 belongs.

1. Counting numbers are {1, 2, 3, . . .}.
   −4 is not a counting number.

2. Whole numbers are {0, 1, 2, 3, . . .}.
   −4 is not a whole number.

3. Integers are {. . ., −3, −2, −1, 0, 1, 2, 3, . . .}.
   −4 is an integer.

4. Rational numbers can be written as fractions.

   $-4 = \frac{-4}{1}$

   −4 is a rational number.

5. −4 is an integer and a rational number.

YOUR TURN!

Identify all sets to which the number $\frac{30}{2}$ belongs.

1. Counting numbers are {1, 2, 3, . . .}.
   $\frac{30}{2}$ simplifies to ______, so it ________ a counting number.

2. Whole numbers are {0, 1, 2, 3, . . .}.
   $\frac{30}{2}$ ________ a whole number.

3. Integers are {. . ., −3, −2, −1, 0, 1, 2, 3, . . .}.
   $\frac{30}{2}$ ________ an integer.

4. Rational numbers can be written as fractions.
   $\frac{30}{2}$ ________ a rational number.

5. $\frac{30}{2}$ is ____________________________________

____

____

Guided Practice

Write each number as a fraction in simplest form.

1. 25

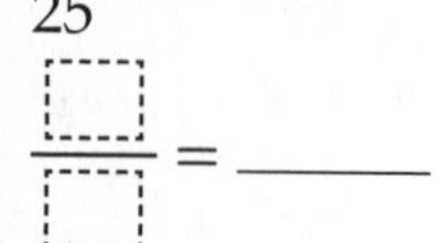

$\frac{\square}{\square}$ = ______

2. −51

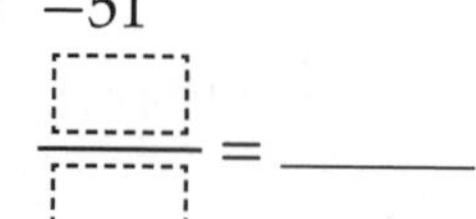

$\frac{\square}{\square}$ = ______

3. −0.56

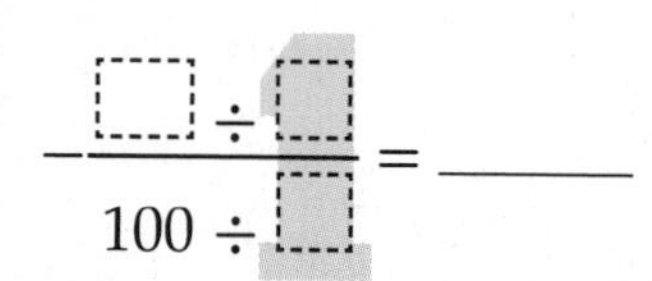

$-\frac{\square \div \square}{100 \div \square}$ = ______

4. 0.40

$\frac{\square \div \square}{100 \div \square}$ = ______

Step by Step Practice

5. Identify all sets to which the number −18 belongs.

Step 1 Counting numbers are {1, 2, 3, . . .}.

−18 ________ a counting number.

Step 2 Whole numbers are {0, 1, 2, 3, . . .}.

−18 ________ a whole number.

Step 3 Integers are {. . ., −3, −2, −1, 0, 1, 2, 3, . . .}.

−18 ________ an integer.

Step 4 Rational numbers can be written as a fraction.

−18 = ________

−18 ________ a rational number.

Step 5 −18 is ________________________________.

Identify all sets to which each number belongs.

6. 5

7. 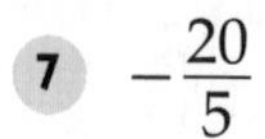 $-\frac{20}{5}$

8. $-3.\overline{7}$

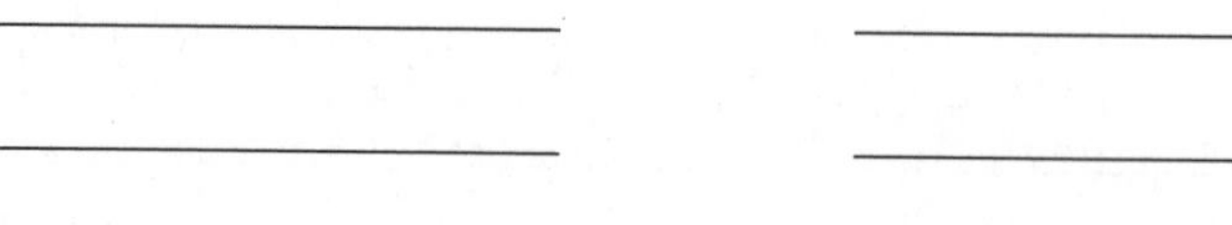

________ ________ ________

________ ________ ________

________ ________ ________

________ ________ ________

Step by Step Problem-Solving Practice

Solve.

9 TRANSPORTATION At Highland Heights High School, $\frac{47}{100}$ students drive or ride in a car to school, one third ride a bus, and 1 out of 5 walk. Order the types of transportation students use to get to school from least to greatest.

car: $\frac{47}{100}$ = ______

bus: one-third = $\frac{1}{3}$ = ______

walk: 1 out of 5 = $\frac{1}{5}$ = ______

Check off each step.

______ **Understand: I underlined key words.**

______ **Plan: To solve the problem, I will** ______________________.

______ **Solve: The answer is** ______________________.

______ **Check: I checked my answer by** ______________________.

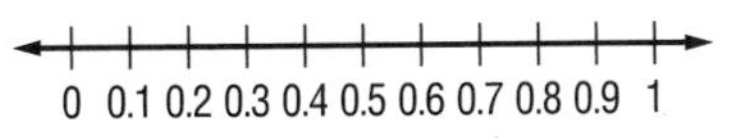

Skills, Concepts, and Problem Solving

Write each number as a fraction in simplest form.

10 −6.2 = ______

11 0.41 = ______

12 0.7 = ______

13 5 = ______

14 55 = ______

15 1.12 = ______

GO ON

Identify all sets to which each number belongs.

16 67

____

____

____

____

17 $4.\overline{6}$

____

____

____

____

18 $-\frac{9}{3}$

____

____

____

____

Solve.

19 **WEATHER** The average snowfall received in January was 1.2 in. Write as a fraction in simplest form.

____

20 **BASKETBALL** Curtis' scoring statistics for the first 5 games of the season are shown in the table below. Name the set(s) that Curtis' score from game 4 belongs to.

Game	1	2	3	4	5
Scoring statistics	0.64	$\frac{4}{9}$	0.39	8 out of 20	$\frac{71}{100}$

____

Vocabulary Check **Write the vocabulary word that completes each sentence.**

21 0.125 is an example of a(n) _______________.

22 −7, 3, 0, 4, 92 are all examples of rational numbers and

_______________.

23 Fractions, terminating and repeating decimals, whole numbers, and integers are all examples of _______________.

24 **Reflect** Explain why one of the numbers below does not belong.

$0.\overline{2}$ 4 out of 12 $\frac{75}{100}$ 0.353553555 . . .

____

____

STOP

Lesson 4-2

Fractions and Decimals

KEY Concept

Fractions and **decimals** are different ways to show the same number.

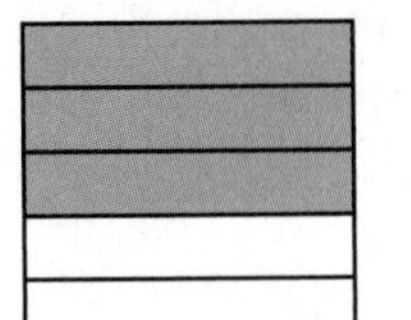 $\frac{3}{5} = 0.6$ 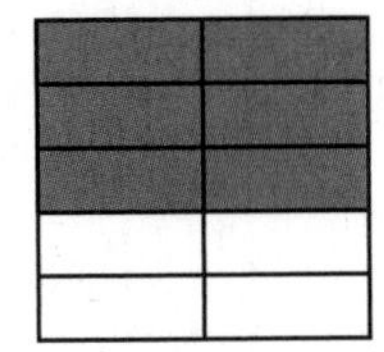

You can convert fractions to decimals and decimals to fractions.

fraction → decimal

Divide the **numerator** by the **denominator**.

"3 divided by 5"

$$\frac{3}{5} = 3 \div 5 \rightarrow 5\overline{)3.0} = 0.6$$

$$\begin{array}{r} 0.6 \\ 5\overline{)3.0} \\ 30 \\ \hline 0 \end{array}$$

decimal → fraction

The numerator is the number without the decimal point. The denominator is a multiple of 10. For each decimal place in the number, there is one 0 in the denominator.

Ask "By what number can I divide both 6 and 10?"

$$0.6 = \frac{6}{10} = \frac{6 \div 2}{10 \div 2} = \frac{3}{5}$$

VOCABULARY

decimal
a number that can represent whole numbers and fractions; a decimal point separates the whole number from the fraction

denominator
the number below the bar in a fraction that tells how many equal parts in the whole or the set

fraction
a number that represents part of a whole or part of a set

numerator
the number above the bar in a fraction that tells how many equal parts are being used

To simplify a fraction, write the fraction so there are no numbers that will divide evenly into both the numerator and denominator.

Example 1

Divide to write $\frac{3}{4}$ as a decimal.

1. Divide 3 by 4.
2. $\frac{3}{4} = 0.75$

$$3 \div 4 \rightarrow \begin{array}{r} 0.75 \\ 4\overline{)3.00} \\ -28 \\ \hline 20 \\ -20 \\ \hline 0 \end{array}$$

YOUR TURN!

Divide to write $\frac{13}{25}$ as a decimal.

1. Divide ______ by ______.
2. $\frac{13}{25}$ = ______ ____ ÷ ____ → [ ] $\overline{)}$ [ ]

−______

−______

Example 2

Write 0.12 as a fraction in simplest form.

1. The numerator will be 12.
2. The denominator will be 100 because the decimal has 2 digits to theright of the decimal point.

$\frac{12}{100}$

3. Divide the numerator and the denominator by their GCF.

Ask "What is the greatest number that divides evenly into both 12 and 100?"

$\frac{12 \div 4}{100 \div 4} = \frac{3}{25}$

YOUR TURN!

Write 0.42 as a fraction in simplest form.

1. The numerator will be ______.
2. The denominator will be ______ because the decimal has ______ digits to the right of the decimal point.

______

3. Divide the numerator and the denominator by their GCF.

$\frac{\square \div \square}{\square \div \square} =$ ______

Guided Practice

Divide to write each fraction as a decimal.

1. $\frac{11}{50} =$ ______ $50\overline{)11.00}$

2. $\frac{6}{15} =$ ______ $15\overline{)6.00}$

3. $\frac{7}{20} =$ ______ $20\overline{)7.00}$

4. $\frac{3}{12} =$ ______ $12\overline{)3.00}$

Step by Step Practice

5 Write 3.65 as a fraction in simplest form.

Step 1 The numerator will be ______.

Step 2 The denominator will be ______ because there are ______ digits to the right of the decimal point.

Step 3 Divide the numerator and the denominator by their GCF.

$$\frac{\square \div \square}{\square \div \square} = \underline{\quad\quad} = \underline{\quad\quad}$$

Write each decimal as a fraction in simplest form.

6 0.08 =

$$\frac{8 \div \square}{100 \div \square} = \underline{\quad\quad}$$

7 0.34 =

$$\frac{34 \div \square}{100 \div \square} = \underline{\quad\quad}$$

8 0.72

$$\frac{\square \div \square}{\square \div \square} = \underline{\quad\quad}$$

9 2.40

$$\frac{\square \div \square}{\square \div \square} = \underline{\quad\quad}$$

10 1.08

$$\frac{\square \div \square}{\square \div \square} = \underline{\quad\quad}$$

11 8.4

$$\frac{84 \div \square}{10 \div \square} = \underline{\quad\quad}$$

GO ON

Step by Step Problem-Solving Practice

Solve.

12 **HOMEWORK** Reid has completed 6 out of 18 problems assigned for homework. Write the homework Reid has completed as a decimal.

$$\frac{6}{18} \rightarrow 18\overline{)6.00}$$

Check off each step.

_______ **Understand: I underlined key words.**

_______ **Plan: To solve the problem, I will** _______________.

_______ **Solve: The answer is** _______________.

_______ **Check: I checked my answer by** _______________.

Skills, Concepts, and Problem Solving

Write each fraction as a decimal.

13 $\frac{5}{8} =$ _______

14 $\frac{13}{20} =$ _______

15 $\frac{2}{4} =$ _______

16 $\frac{1}{9} =$ _______

17 $\frac{1}{20} =$ _______

18 $\frac{12}{25} =$ _______

Write each decimal as a fraction in simplest form.

19 0.6 = ________ **20** 0.07 = ________ **21** 1.59 = ________

22 $0.\overline{3}$ = ________ **23** 2.9 = ________ **24** 0.054 = ________

Solve.

25 **SURVEY** Mr. Vincent took a class survey and found that 11 out of 22 students would rather listen to music than watch television. Write the number of students who chose music as both a fraction and a decimal.

____

26 **MONEY** Lyla had $12 in her pocket. While walking home from school she found two quarters and two dimes on the sidewalk. Write the amount of money she has both as a fraction and a decimal.

____

27 **FUNDRAISING** Of the 310 sophomore students at Johnston High School, 248 of them participated in the school's annual fundraiser. Write the number of sophomores that participated as both a fraction and a decimal.

____

Vocabulary Check **Write the vocabulary word that completes each sentence.**

28 The number above the bar in a fraction is the ____________.

29 The number below the bar in a fraction is the ____________.

30 A number that represents part of a whole or part of a set is known as a(n) ____________.

31 The numbers 5.2, 0.19, and 0.3572 are all examples of a(n) ____________.

32 **Reflect** When you are given a decimal to write as a fraction, how do you know if the fraction will be an improper fraction?

____

____

____

STOP

Chapter 4

Progress Check 1 (Lessons 4-1 and 4-2)

Write each number as a fraction in simplest form.

1 $0.76 =$ ________

2 6 out of 10 = ________

3 $-0.3 =$ ________

4 $2.75 =$ ________

5 $14 =$ ________

6 $3.36 =$ ________

Write each fraction as a decimal.

7 $\frac{1}{8} =$ ________

8 $\frac{4}{9} =$ ________

9 $\frac{9}{12} =$ ________

10 $\frac{5}{9} =$ ________

Write each decimal as a fraction in simplest form.

11 $0.44 =$ ________

12 $0.09 =$ ________

13 $0.83 =$ ________

14 $6.4 =$ ________

15 $0.12 =$ ________

16 $3.71 =$ ________

Solve.

17 **PRECIPITATION** The average rainfall received in July was 6.4 mm. Write as a mixed number in simplest form.

____

18 **TRAVEL** Phil drove 3 hours and 48 minutes to his grandmother's house. Write the time as a fraction and a decimal.

____

Lesson 4-3

Decimals and Percents

KEY Concept

A **percent** is a ratio that compares a number to 100. Percent means "per hundred" or "out of one hundred."

Decimals and percents are different ways to show the same value.

25 hundredths $= \frac{25}{100} = 0.25 = 25\%$

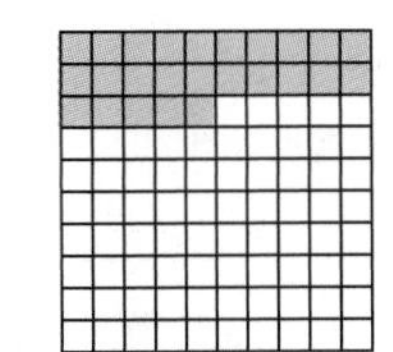

decimal → percent

Multiply the decimal by 100 and write a percent symbol.

To multiply by 100, move the decimal point right two places.

$0.25 \cdot 100 = 25\%$

percent → decimal

Remove the percent symbol and divide the decimal by 100.

To divide by 100, move the decimal point left two places.

$25 \div 100 = 0.25$

VOCABULARY

decimal
a number that can represent whole numbers and fractions; a decimal point separates the whole number from the fraction

percent
a ratio that compares a number to 100

ratio
a comparison of two numbers by division

Percents are used in many real-world applications such as statistics and weather predictions.

Example 1

What percent of the model is shaded?

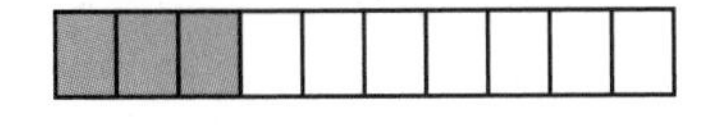

1. The model has 10 equal parts.
   3 of the parts are shaded.
2. The ratio for the shaded parts is $\frac{3}{10}$.
3. Convert the fraction to a decimal.

   $\frac{3}{10} = 0.3$
4. Convert the decimal to a percent.

   $0.3 \cdot 100 = 30\%$

YOUR TURN!

What percent of the model is shaded?

1. The model has _____ equal parts.
   _____ of the parts are shaded.
2. The ratio for the shaded parts is _____.
3. Convert the fraction to a decimal.

   _____ = _____
4. Convert the decimal to a percent.

   _____ • 100 = _____

GO ON

Example 2

Write 0.07 as a percent.

1. Multiply the decimal by 100.

   0.07 • 100 = 7

2. Write a percent symbol.

   7%

YOUR TURN!

Write 1.15 as a percent.

1. ________ the decimal by ________.

   1.15 • 100 = ________

2. Write a percent symbol.

   ____

Example 3

Write 175% as a decimal.

1. Remove the percent symbol.
2. Divide the number by 100.

   175 ÷ 100 = 1.75

YOUR TURN!

Write 9% as a decimal.

1. Remove the percent symbol.
2. ________ the number by ____.

   9 ÷ 100 = ________

Guided Practice

What percent of each model is shaded?

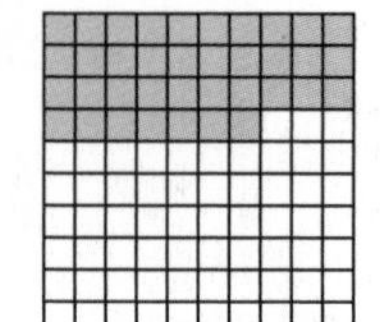

fraction: ______

decimal: ______

percent: ______

fraction: ______

decimal: ______

percent: ______

Write each decimal as a percent.

3 0.73

0.73 • ______ = ______

______%

4 3.6

3.6 • ______ = ______

______%

Step by Step Practice

5 Write 16.25% as a decimal.

Step 1 ________ the percent symbol.

Step 2 ________ by ________.

16.25 ÷ ________ = ________

Write each percent as a decimal.

6 60%

______ ÷ 100 = ______

7 22.5%

______ ÷ 100 = ______

8 2%

______ ÷ 100 = ______

9 146%

______ ÷ 100 = ______

Step by Step Problem-Solving Practice

Solve.

10 **JOGGING** Tammy and Cameron were jogging on a 1-mile circuit. Cameron had to stop to tie his shoe after they had run 0.55 mile. What percent of the circuit had they finished?

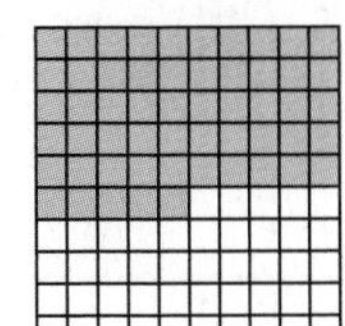

0.55 = ______ = ______

Check off each step.

______ **Understand: I underlined key words.**

______ **Plan: To solve the problem, I will** ________________________________.

______ **Solve: The answer is** ________________________________.

______ **Check: I checked my answer by** ________________________________.

GO ON

Skills, Concepts, and Problem Solving

Write each decimal as a percent.

11 0.15 = ________ 12 1.5 = ________ 13 0.015 = ________

14 2.2 = ________ 15 0.06 = ________ 16 0.0045 = ________

Write each percent as a decimal.

17 30% = ________ 18 3% = ________ 19 805% = ________

20 62.5% = ________ 21 66% = ________ 22 200% = ________

Solve.

23 **BASEBALL** Kurt's batting average for the season is 0.642. Batting average is a ratio of hits to at-bats. What percent of at-bats did Kurt get a hit?

____

24 **BUDGETS** The circle graph at the right illustrates how the Fisher family spends their annual income. Write the percent of income that is spent on utilities as a decimal.

____

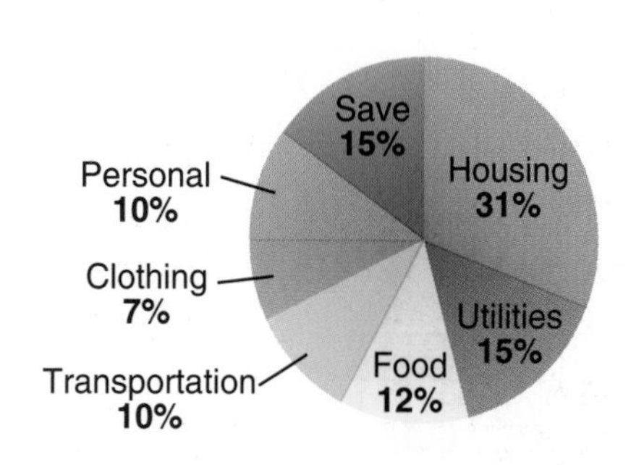

Vocabulary Check **Write the vocabulary word that completes each sentence.**

25 A(n) ________________ is a comparison of two numbers by division.

26 ________________ means "per hundred."

27 A(n) ________________ separates the whole number from the fraction.

28 **Reflect** Rocco was asked to write a decimal that is between 0 and 1 and also display it as a percent. He wrote 0.5 and 5%. Is he correct? Explain.

____

____

STOP

Lesson 4-4

Fractions and Percents

KEY Concept

Fractions and **percents** are different ways to show the same value. You can convert between them using different methods.

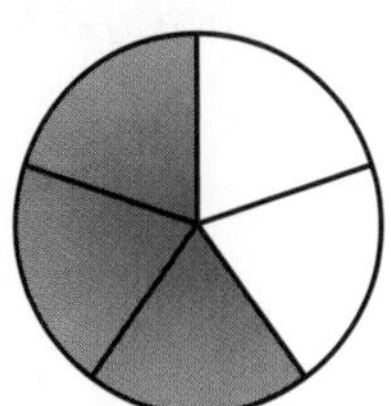

$$\frac{3}{5} = \frac{3 \cdot 20}{5 \cdot 20} = \frac{60}{100} = 60\%$$

Method 1

fraction → percent — Write and solve a **proportion**.

$\frac{3}{5} = \frac{x}{100}$ — 3 out of 5 is the same as x out of 100.

$3 \cdot 100 = 5 \cdot x$ — Cross multiply.

$300 = 5 \cdot x$ — What number times 5 equals 300?

$\frac{300}{5} = x$ — Divide 300 by 5.

$60 = x$

$\frac{3}{5} = 60\%$

Method 2

fraction → decimal → percent

Divide the numerator by the denominator to change the fraction to a decimal. Then multiply by 100 to change the decimal to a percent.

$$\frac{3}{5} = 3 \div 5 \rightarrow 5\overline{)3.00} = 0.60$$

$$0.60 \cdot 100 = 60\%$$

percent → decimal → fraction

Remove the percent symbol. Write as a fraction with 100 in the denominator. Simplify.

$$60\% \rightarrow 0.60 \rightarrow \frac{60}{100} = \frac{60 \div 20}{100 \div 20} = \frac{3}{5}$$

VOCABULARY

denominator
the number below the bar in a fraction that tells how many equal parts in the whole or the set

fraction
a number that represents part of a whole or set

percent
a ratio that compares a number to 100

proportion
an equation stating that two ratios are equivalent

Commonly used fractions and percents are $25\% = \frac{1}{4}$, $50\% = \frac{1}{2}$, and $75\% = \frac{3}{4}$.

GO ON

Example 1

Write the percent that names the shaded part of the model.

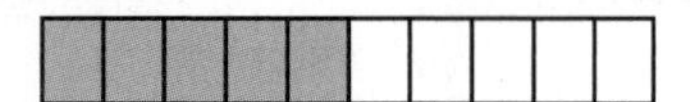

1. Write the fraction that represents the model.
   $\frac{5}{10}$

2. Write a proportion.
   $\frac{5}{10} = \frac{x}{100}$

3. Solve for x.

   $5 \cdot 100 = 10 \cdot x$

   $500 = 10 \cdot x$

   $\frac{500}{10} = x$

   $50 = x$

4. Write as a percent.
   $\frac{5}{10} = \frac{50}{100} = 50\%$

YOUR TURN!

Write the percent that names the shaded part of the model.

1. Write the fraction that represents the model.

   ______

2. Write a proportion.

   ______ = ______

3. Solve for x.

   ______ • ______ = ______ • ______

   ______ = ______ • ______

   ______ = ______

   ______ = ______

4. Write as a percent.

   ______ = ______ = ______

Example 2

Write 225% as a fraction in simplest form.

1. Remove the percent symbol. Write the number as a fraction with 100 in the denominator.

   $225\% = \frac{225}{100}$

2. Simplify. 225 and 100 have a GCF of 25.

   $\frac{225 \div 25}{100 \div 25} = \frac{9}{4} = 2\frac{1}{4}$

   Divide the numerator and denominator by the GCF.

YOUR TURN!

Write 165% as a fraction in simplest form.

1. Remove the percent symbol. Write the number as a fraction with 100 in the denominator.

   ______ = $\frac{\square}{100}$

2. Simplify. 165 and 100 have a GCF of ______.

   $\frac{\square \div \square}{100 \div \square}$ = ______ = ______

Example 3

Write $\frac{8}{20}$ as a percent by dividing.

1. Divide the numerator by the denominator.

$$\begin{array}{r} 0.4 \\ 20\overline{)8.0} \\ -\,80 \\ 0 \end{array}$$

2. Write the decimal as a percent.

$0.40 \cdot 100 = 40\%$

YOUR TURN!

Write $\frac{12}{5}$ as a percent by dividing.

1. Divide the numerator by the denominator.

$$\begin{array}{r} 5\overline{)\ \ \ 0} \\ - \underline{\quad} \\ - \underline{\quad} \\ 0 \end{array}$$

2. Write the decimal as a percent.

_____ $\cdot$ 100 = _____%

Guided Practice

Write the percent that names the shaded part of each model.

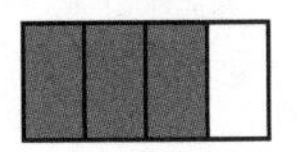

$\frac{\square}{\square} = \frac{x}{\square}$

_____ $\cdot$ _____ = _____ $\cdot\, x$

$\frac{\square}{\square} = x$

_____ $= x$

_____ = _____

2 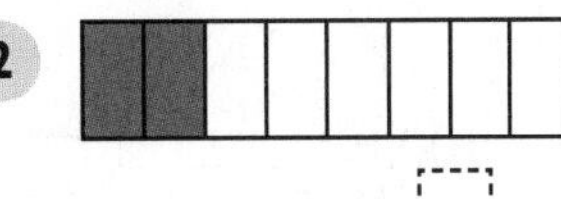

$\frac{\square}{\square} = \frac{x}{\square}$

_____ $\cdot$ _____ = _____ $\cdot\, x$

$\frac{\square}{\square} = x$

_____ $= x$

_____ = _____

Write each percent as a fraction in simplest form.

3 $84\% = \frac{\square}{100}$

84 and 100 have a GCF of _____.

$\frac{\square \div \square}{100 \div \square} =$ _____

4 $170\% = \frac{\square}{100}$

170 and 100 have a GCF of _____.

$\frac{\square \div \square}{100 \div \square} =$ _____

GO ON

Step by Step Practice

5 Write $\frac{7}{8}$ as a percent.

$8\overline{)7.000}$

Step 1 Divide the numerator by the denominator.

Step 2 Write the decimal as a percent.

$\_\_\_\_\_\_ \cdot 100 = \_\_\_\_\_\_\%$

Write each fraction as a percent.

6 $\frac{3}{12}$

$12\overline{)3.00}$ $\_\_\_\_\_\_ \cdot 100 = \_\_\_\_\_\_$

7 $\frac{9}{20}$

$20\overline{)9.00}$ $\_\_\_\_\_\_ \cdot 100 = \_\_\_\_\_\_$

Step by Step Problem-Solving Practice

Solve.

8 **MUSICAL** In the freshman class, 143 out of the 260 students attended a school musical. What percent of the freshman class attended the musical?

$260\overline{)143.00}$

$\_\_\_\_\_\_ \cdot 100 = \_\_\_\_\_\_$

Check off each step.

______ **Understand: I underlined key words.**

______ **Plan: To solve the problem, I will** ________________________.

______ **Solve: The answer is** ________________________.

______ **Check: I checked my answer by** ________________________.

Skills, Concepts, and Problem Solving

Write each fraction as a percent.

9 $\frac{15}{10}$ = ________

10 $\frac{1}{10}$ = ________

11 $\frac{1}{4}$ = ________

12 $\frac{1}{8}$ = ________

13 $\frac{3}{16}$ = ________

14 $\frac{22}{25}$ = ________

Write each percent as a fraction in simplest form.

15 50% = ________

16 34% = ________

17 26% = ________

18 90% = ________

19 6% = ________

20 87% = ________

Solve.

21 **WEATHER** It snowed 14 days in the month of February. What percent of days did it snow in the month of February?

____

February 2010						
Su	M	Tu	W	Th	F	Sa
	1	2	3	4	5	6
7	8	9	10	11	12	13
14	15	16	17	18	19	20
21	22	23	24	25	26	27
28						

22 **FINANCIAL LITERACY** Audon wants to purchase a sweater that is 40% off the original price. Write the discount percent as a fraction.

____

23 **TESTS** Frances scored 36 out of 45 points on her science test. Write her score as a percent.

____

Vocabulary Check **Write the vocabulary word that completes each sentence.**

24 A(n) ________ is represented by the symbol %.

25 The number below the bar in a fraction is the ________.

26 **Reflect** Rebecca was asked to write four fractions that are equivalent to 40%. Her results are shown. Did she complete the task correctly? Explain.

____

____

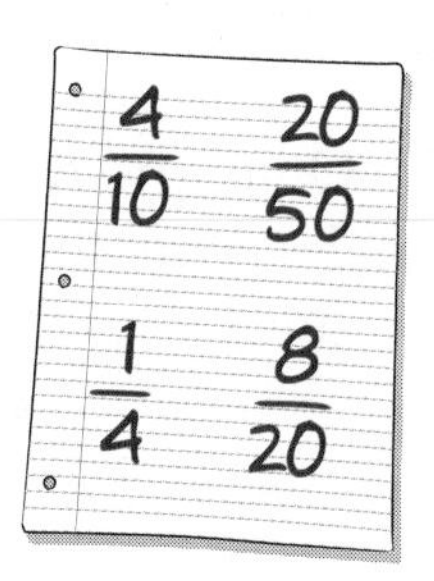

Chapter 4

Progress Check 2 (Lessons 4-3 and 4-4)

Write each decimal as a percent.

1. 0.37 = ______
2. 0.45 = ______
3. 0.079 = ______

Write each percent as a decimal.

4. 600% = ______
5. 18% = ______
6. 20% = ______

Write each percent as a fraction in simplest form.

7. 66% = ______
8. 20% = ______
9. 17% = ______

Write each fraction as a percent.

10. $\frac{15}{25}$ = ______
11. $\frac{1}{10}$ = ______
12. $\frac{46}{50}$ = ______

Solve.

13. **GARDENING** The pie graph to the right shows the types of vegetables Mrs. Washington grows in her garden. Written as a decimal, what percent of her garden is tomatoes?

____

Mrs. Washington's Vegetables

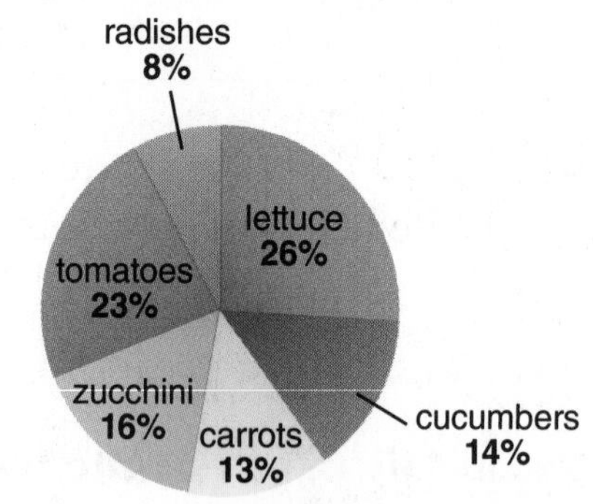

14. **PRECIPITATION** It rained for 6 out of the last 14 days. Write the number of days it has rained as a fraction.

____

15. **SCIENCE** Of all the 125 animals at the zoo, 60 of the animals are mammals. Write the number of mammals as a percent.

____

Lesson 4-5

Squares and Square Roots

KEY Concept

A **perfect square** is the product of a number multiplied by itself. To model a **perfect square**, you can use algebra tiles.

1	1	1	1	1	1
1	1	1	1	1	1
1	1	1	1	1	1
1	1	1	1	1	1
1	1	1	1	1	1
1	1	1	1	1	1

The length and width are each 6 tiles.
$6 \cdot 6 = 6^2$

The square root is the length of one side. So, $\sqrt{36}$ is 6.

Every positive number has a positive and a negative square root.

$$(-6) \cdot (-6) = 36 \text{ and } 6 \cdot 6 = 36$$

$\sqrt{36}$ represents the positive square root of 36.

$$\sqrt{36} = 6$$

$-\sqrt{36}$ represents the negative square root of 36.

$$-\sqrt{36} = -6$$

$\pm\sqrt{36}$ represents both the positive and negative square root of 36.

$$\pm\sqrt{36} = \pm 6$$

A negative number does not have a square root because two equal factors cannot have a negative product.

VOCABULARY

irrational number
a number that cannot be expressed as $\frac{a}{b}$, where a and b are integers and $b \neq 0$

perfect square
a rational number whose square root is a whole number

radical sign
the symbol used to indicate a nonnegative square root, $\sqrt{\ }$

rational number
numbers of the form $\frac{a}{b}$, where a and b are integers and $b \neq 0$

square root
one of two equal factors of a number

When a number under the radical sign is not a perfect square, it is an irrational number and can be estimated.

Example 1

Use a model to find $\sqrt{9}$.

1. $\sqrt{9}$ means the positive square root of 9.
2. Use 9 algebra tiles to make a square.

1	1	1
1	1	1
1	1	1

3. There are 3 tiles on each side of the square. So, $\sqrt{9} = 3$.

YOUR TURN!

Use a model to find $\sqrt{4}$.

1. $\sqrt{4}$ means the ________ square root of ________.
2. Use ________ algebra tiles to make a square.
3. There are ________ tiles on each side of the square. So, $\sqrt{4}$ = ________.

GO ON

Example 2

Find $-\sqrt{49}$.

1. $-\sqrt{49}$ means the negative square root of 49.
2. Find the square root of 49.

   $49 = 7 \cdot 7 = 7^2$

   $\sqrt{49} = 7$
3. So, $-\sqrt{49} = -7$.

YOUR TURN!

Simplify $-\sqrt{64}$.

1. $-\sqrt{64}$ means the ________ square root of ______.
2. Find the square root of ______.

   $64 =$ ______ $\cdot$ ______ $=$ ______

   $\sqrt{64} =$ ______
3. So, $-\sqrt{64} =$ ______.

Example 3

Determine if $\sqrt{6}$ is a rational or irrational number. If rational, find the square root.

1. 6 is not a perfect square.
2. So, $\sqrt{6}$ is an irrational number.

YOUR TURN!

Determine if $\sqrt{36}$ is a rational or irrational number. If rational, find the square root.

1. 36 ________ a perfect square.
2. So, $\sqrt{36}$ is ________ number.

   $\sqrt{36} =$ ________

Guided Practice

Use a model to find each square root.

1. $\sqrt{16} =$ ________
2. $\sqrt{25} =$ ________
3. $\sqrt{81} =$ ________
4. $\sqrt{64} =$ ________

Step by Step Practice

5 Determine if $\pm\sqrt{121}$ is a rational or irrational number. If rational, find the square root.

Step 1 121 ________ a perfect square.

Step 2 So, $\sqrt{121}$ is a __________ number.

Step 3 The $\pm$ symbol in $\pm\sqrt{121}$ means both the

____________________________________ of 121.

Step 4 ________ • ________ or ________ = ________

So, $\pm\sqrt{121}$ = ________

Determine if each number is a rational or irrational number. If rational, find the square root.

6 $\sqrt{100}$ = ________

100 = ________ • ________

7 $-\sqrt{144}$ = ________

144 = ________ • ________

Step by Step Problem-Solving Practice

Solve.

8 **CONSTRUCTION** Randy is building a square sandbox for his younger sister. The area of the sandbox is to be 36 ft^2. If they are using wooden boards for the sides of the sandbox, how many feet of boards will they need?

$\sqrt{36}$ = _____

Each side will be _____.
The perimeter of the sandbox is _____ + _____ + _____ + _____ = _____.

Check off each step.

______ **Understand: I underlined key words.**

______ **Plan: To solve the problem, I will** ____________________.

______ **Solve: The answer is** ____________________.

______ **Check: I checked my answer by** ____________________.

GO ON

Skills, Concepts, and Problem Solving

Find each square root.

9 $-\sqrt{64}$ = ____________

10 $\pm\sqrt{196}$ = ____________

11 $\pm\sqrt{225}$ = ____________

12 $\sqrt{400}$ = ____________

Determine if each number is a rational or irrational number. If rational, find the square root.

13 $\sqrt{39}$ ____________

14 $-\sqrt{110}$ ____________

15 $\sqrt{169}$ ____________

16 $\pm\sqrt{1}$ ____________

17 $\sqrt{1600}$ ____________

18 $-\sqrt{55}$ ____________

Solve.

19 **SIGNS** A hospital road sign is a perfect square. The sign's side length is 24 inches. What is the area of the sign?

____

20 **BASEBALL** An infield of a baseball diamond is a perfect square. The distance from first base to second base is 90 ft. What is the area of the baseball diamond?

____

Vocabulary Check **Write the vocabulary word that completes each sentence.**

21 A(n) ____________ is the product of a number multiplied by itself.

22 One of two equal factors of a number is called a(n)

____________.

23 Any number that can not be written as a fraction is known as a(n)

____________.

24 **Reflect** Juan is hoping to create a square bulletin board in his dorm room. The pieces of cork each have an area of one square foot. If he has 46 pieces of cork, will he be able to create a square bulletin board? If so, what will be the dimensions of the largest bulletin board he can make?

____

____

STOP

Lesson 4-6

Compare and Order Real Numbers

KEY Concept

The set of **real numbers** is all the **rational** and the **irrational** numbers.

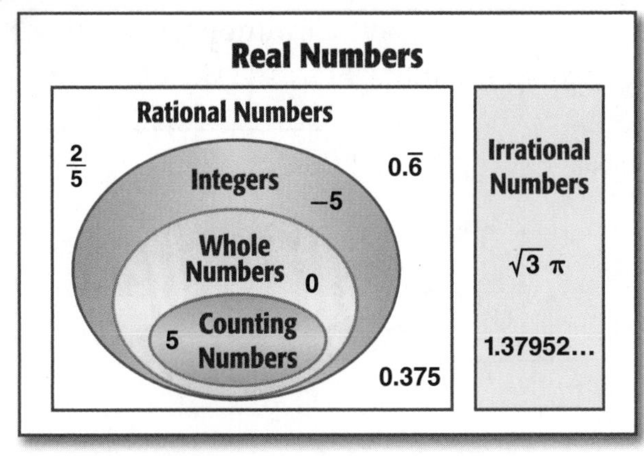

To compare real numbers, you can use a number line.

Use >, <, or = to compare numbers.

$-\sqrt{4}$, 0.75, $\sqrt{16}$ on a number line from −4 to 4

$-\sqrt{4} < 0.75 < \sqrt{16}$

The larger numbers are farther to the right.

VOCABULARY

irrational number
a number that cannot be expressed as $\frac{a}{b}$, where a and b are integers and $b \neq 0$

rational number
numbers of the form $\frac{a}{b}$, where a and b are integers and $b \neq 0$

real numbers
the set of rational numbers together with the set of irrational numbers

To compare numbers, write each number as a decimal. Then compare the decimals. If the number is irrational, estimate to find a decimal.

Example 1

Identify all sets to which the number $\sqrt{3}$ belongs.

1. Determine if $\sqrt{3}$ is a rational or irrational number.

   Because 3 is not a perfect square, $\sqrt{3}$ is an irrational number.

2. Because $\sqrt{3}$ is an irrational number it cannot belong to any subsets of the rational numbers.

3. So, $\sqrt{3}$ belongs to the set of real numbers and the set of irrational numbers.

YOUR TURN!

Identify all sets to which the number $-\sqrt{49}$ belongs.

1. Determine if $-\sqrt{49}$ is a rational or irrational number.

   Because 49 ________ a perfect square, $-\sqrt{49}$ is ________________ number.

2. Because $-\sqrt{49}$ is ________________ number it ________________ to subsets of the rational numbers.

3. So, $-\sqrt{49}$ belongs to ____________ ________________________ ________________________ ________________.

GO ON

Example 2

Write >, <, or = to compare $-\sqrt{81}$ and $-\frac{31}{3}$.

1. Write each number as a decimal.

$-\sqrt{81} = -9.0 \qquad -\frac{31}{3} = -10.\overline{3}$

2. Graph the numbers on a number line.

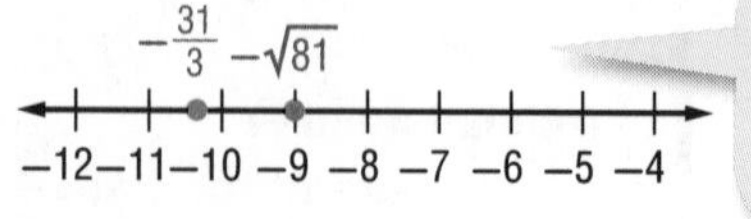

$-\sqrt{81}$ is on the right so it is the greater number.

3. Write an inequality.

$-\sqrt{81} > -\frac{31}{3}$

YOUR TURN!T

Write >, <, or = to compare 0.3 and $\frac{2}{3}$.

1. Write each number as a decimal.

$0.\overline{3}$ = ________ $\frac{2}{3}$ = ________

2. Graph the numbers on a number line.

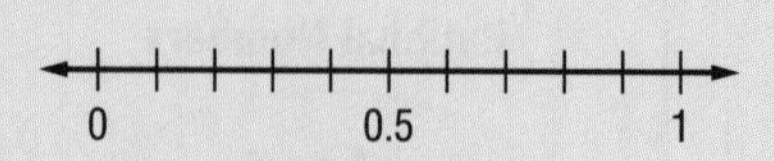

3. Write an inequality.

$0.\overline{3} \bigcirc \frac{2}{3}$

Guided Practice

Identify all sets to which each number belongs.

1. $-\sqrt{64}$

____

____

____

2. 0.256184...

____

____

____

Step by Step Practice

3. Write >, <, or = to compare $-7\frac{1}{4}$ and $-\frac{45}{6}$.

Step 1 Write each number as a decimal.

$-7\frac{1}{4}$ = ________ $-\frac{45}{6}$ = ________

Step 2 Graph the numbers on a number line.

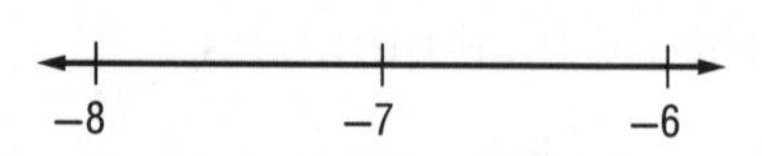

Step 3 Write the inequality. $-7\frac{1}{4} \bigcirc -\frac{45}{6}$

Write >, <, or = to compare the numbers.

4 $-\sqrt{169}$ ◯ -12.75

$-\sqrt{169} =$ ______

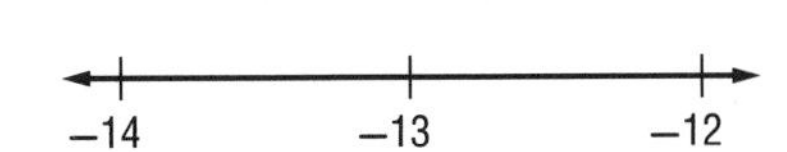

5 $10\frac{5}{6}$ ◯ $10.8\overline{3}$

$10\frac{5}{6} =$ ______

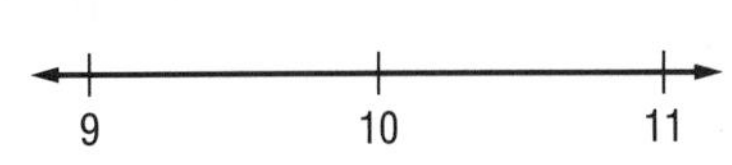

Step by Step Problem-Solving Practice

Solve.

6 **CONTEST** Jose is in a math contest. He will be the champion if he can correctly order the times in the table from least to greatest. In what order should the times be?

Time (hr)	Decimal
2.85	______
$2\frac{3}{5}$	______
2.3	______
$\sqrt{4}$	______

Check off each step.

______ **Understand: I underlined key words.**

______ **Plan: To solve the problem, I will** ______.

______ **Solve: The answer is** ______.

______ **Check: I checked my answer by** ______.

Skills, Concepts, and Problem Solving

Identify all sets to which each number belongs.

7 -95

______
______
______

8 $\frac{5}{8}$

______
______
______

9 14π

______
______
______

GO ON

Write >, <, or = to compare the numbers.

10 $1.5 \bigcirc \frac{1}{5}$

11 $-5.2 \bigcirc -5\frac{3}{4}$

12 $\sqrt{18} \bigcirc 4.9$

13 $2\frac{5}{10} \bigcirc \sqrt{25}$

14 $6.2 \bigcirc 6\frac{1}{5}$

15 $7.0 \bigcirc \sqrt{49}$

16 $\frac{3}{9} \bigcirc \sqrt{9}$

17 $\sqrt{5} \bigcirc 1.3$

18 $\sqrt{64} \bigcirc 7\frac{1}{2}$

19 $-4.8 \bigcirc -4\frac{2}{5}$

20 $2.5 \bigcirc \sqrt{6}$

21 $-4.1 \bigcirc -\sqrt{16}$

Solve.

22 **TRACK** Five members of the Hillcrest track team participated in the 800-meter race. Their times are in the table. Order the racers' times from least to greatest.

____

Runner	Time
Katie	$2\frac{4}{5}$ min
Helena	$\sqrt{9}$ min
Josie	2.87 min
Pam	$2\frac{1}{2}$ min
Rachel	3.05 min

23 **CARS** Tanya is considering a new car. She wants a car that holds more fuel. The car she has now holds 10.2 gallons. The new car holds $10\frac{1}{2}$ gallons. Should she buy a new car?

____

Vocabulary Check **Write the vocabulary word that completes each sentence.**

24 Any number that cannot be written as a fraction is a(n) ______________.

25 Any number that can be written as a fraction is a(n) ______________.

26 ______________ are made up of both rational and irrational numbers.

27 **Reflect** Although $\sqrt{12}$ is not a perfect square, it is still possible to estimate its value by finding the two closest perfect squares it lies between. Name those two perfect squares and explain how you can estimate the value of $\sqrt{12}$.

____

____

STOP

Chapter 4

Progress Check 3 (Lessons 4-5 and 4-6)

Determine if each number is a rational or irrational number. If rational, find the square root.

1 $-\sqrt{16}$ ____________

2 $\sqrt{30}$ ____________

3 $\pm\sqrt{150}$ ____________

4 $\sqrt{36}$ ____________

5 $\pm\sqrt{144}$ ____________

6 $-\sqrt{220}$ ____________

7 $\sqrt{49}$ ____________

8 $\sqrt{12}$ ____________

9 $-\sqrt{225}$ ____________

Write >, <, or = to compare the numbers.

10 $6.5 \bigcirc \sqrt{49}$

11 $-8.75 \bigcirc -8\frac{2}{3}$

12 $\sqrt{81} \bigcirc 9.5$

13 $4\frac{4}{5} \bigcirc \sqrt{25}$

14 $9.8 \bigcirc 9\frac{4}{25}$

15 $8.0 \bigcirc \sqrt{72}$

16 $\frac{16}{2} \bigcirc \sqrt{64}$

17 $\sqrt{16} \bigcirc 3.5$

18 $\sqrt{100} \bigcirc 5\frac{1}{5}$

Solve.

19 **AREA** A window at the front of a shop is a perfect square. The window's area is 9 m^2. What is the length of each side?

____

20 **ADVERTISEMENT** For a promotion, a car dealership has advertised the lengths of five vehicles as shown in the table. At noon, the first person to write the lengths of the vehicles in order from least to greatest wins a free oil change. What is the correct order of the lengths?

____

Vehicle	Length (feet)
sedan	$10\frac{2}{7}$
pick up	$\sqrt{144}$
compact	8.25
van	$13\frac{1}{4}$
station wagon	10.2

Chapter 4

Chapter Test

Write each decimal as a fraction in simplest form.

1. $0.56 =$ ______
2. $-0.05 =$ ______
3. $0.9 =$ ______
4. $0.06 =$ ______

Write each fraction as a decimal.

5. $\frac{2}{3} =$ ______
6. $\frac{7}{8} =$ ______

Write each decimal as a percent.

7. $0.07 =$ ______
8. $0.013 =$ ______
9. $0.3 =$ ______

Write each percent as a decimal.

10. $55\% =$ ______
11. $3\% =$ ______
12. $61\% =$ ______

Write each fraction as a percent.

13. $\frac{9}{10} =$ ______
14. $\frac{7}{100} =$ ______
15. $\frac{7}{8} =$ ______

Write each percent as a fraction in simplest form.

16. $88\% =$ ______
17. $130\% =$ ______
18. $8\% =$ ______

Determine if each number is a rational or irrational number. If rational, simplify.

19. $-\sqrt{36}$ ______
20. $-\sqrt{64}$ ______
21. $\pm\sqrt{18}$ ______
22. $\sqrt{42}$ ______
23. $\pm\sqrt{25}$ ______
24. $-\sqrt{81}$ ______
25. $\sqrt{100}$ ______
26. $\sqrt{15}$ ______
27. $-\sqrt{169}$ ______

Write $>$, $<$, or $=$ to compare the numbers.

28 $4 \bigcirc \sqrt{16}$

29 $-1.9 \bigcirc -1\frac{1}{2}$

30 $\sqrt{81} \bigcirc 9.5$

31 $3\frac{1}{3} \bigcirc \sqrt{6}$

32 $5.5 \bigcirc 5\frac{5}{10}$

33 $3 \bigcirc \sqrt{15}$

34 $\frac{7}{1} \bigcirc \sqrt{49}$

35 $\sqrt{9} \bigcirc 2$

36 $\sqrt{4} \bigcirc \frac{5}{2}$

Solve.

37 **TIME** Tommy spends 8 hours doing homework each week. He spends 2 of those hours doing language arts homework. Write the fraction and decimal of the time he spends doing language arts.

____

38 **BUDGET** The circle graph shows Tennille's weekly budget. Write the percent of her budget that she spends on clothing as a decimal and a fraction.

____

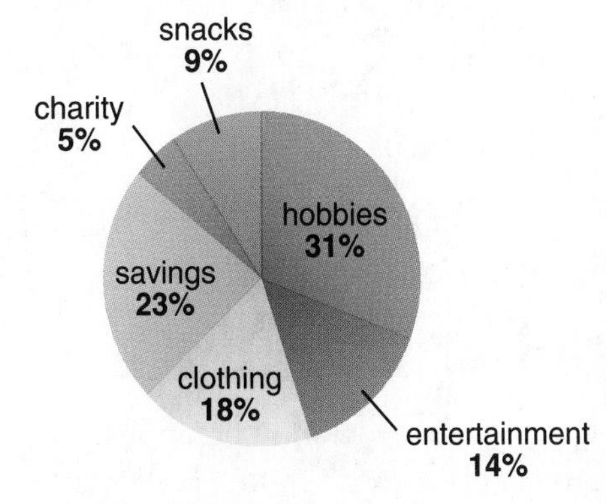

Correct the mistake.

39 The automobile dealership that ran the advertisement shown below had 160 cars on the lot. Of those cars, 20 were red. Is the advertisement accurate? If not, correct the advertisement.

____

____

STOP

Chapter 5

Measurement and Geometry

How much will it cost?

The cost of produce such as fruits and vegetables is often priced by the pound. In a grocery store, you can find how much you will pay for produce by multiplying the weight by the price per pound.

STEP 1 Chapter Pretest Are you ready for Chapter 5? Take the Chapter 5 Pretest to find out.

STEP 2 Preview Get ready for Chapter 5. Review these skills and compare them with what you will learn in this chapter.

What You Know

You know how to evaluate expressions.

Example:

$\frac{7 \cdot 8}{2} = \frac{56}{2}$ Multiply 7 by 8.

$= 28$ Divide 56 by 2.

TRY IT!

Simplify.

1. $6 \cdot 9 =$ ______

2. $12 \cdot 14 =$ ______

3. $\frac{5 \cdot 10}{2} =$ ______

4. $\frac{8 \cdot 9}{2} =$ ______

What You Will Learn

Lesson 5-5

You can find the area of a triangle by multiplying the base by the height and dividing by 2.

Example:

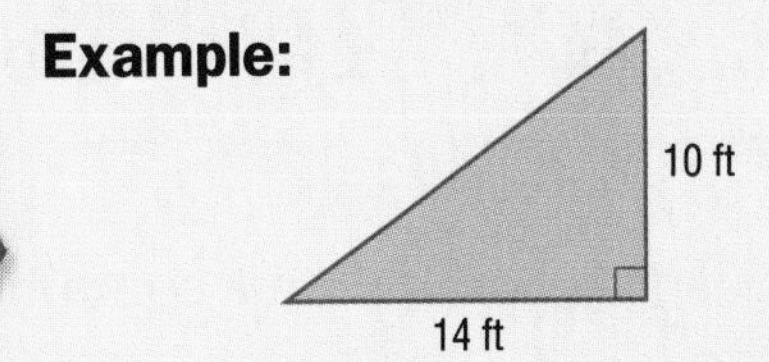

$\frac{14 \cdot 10}{2} = \frac{140}{2}$ Multiply 14 by 10.

$= 70$ Divide 140 by 2.

The area of the triangle is 70 square feet.

What You Know

You know how to simplify expressions.

Examples:

$(2)(3)(4) + (2)(3)(5) + (2)(4)(5)$

$= 24 + 30 + 40$

$= 94$

TRY IT!

Simplify.

5. $4 \cdot 7 \cdot 8 =$ ______

6. $6 \cdot 9 \cdot 11 =$ ______

7. $(2)(6)(7) + (2)(6)(9) =$ ____

8. $(2)(8)(5) + (2)(8)(6) =$ ____

What You Will Learn

Lesson 5-6

The volume V of a rectangular prism is the product of its length ℓ, width w, and height h.

$V = \ell wh$

4 cm, 3 cm, 9 cm

$= (9)(3)(4)$

$= 108 \text{ cm}^3$

Lesson 5-1

The Customary System

KEY Concept

The **customary system** is used in the United States.

Length is a measure of how long something is.

Unit	Abbreviation	Equivalent	Example
inch	in.	——	paperclip
foot	ft	1 ft = 12 in.	notebook
yard	yd	1 yd = 3 ft	baseball bat
mile	mi	1 mi = 5,280 ft	15 football fields

Capacity measures how much a container can hold.

Unit	Abbreviation	Equivalent	Example
fluid ounce	fl oz	——	eye dropper
cup	c	1 c = 8 fl oz	coffee mug
pint	pt	1 pt = 2 c	cereal bowl
quart	qt	1 qt = 2 pt	pitcher
gallon	gal	1 gal = 4 qt	milk jug

The **weight** of an object tells how heavy it is.

Unit	Abbreviation	Equivalent	Example
ounce	oz	——	one strawberry
pound	lb	1 lb = 16 oz	loaf of bread
ton	T	1 T = 2,000 lb	car

To convert between units, multiply or divide using an equivalent measure. When converting to a smaller unit, multiply. When converting to a larger unit, divide.

Another way to convert between units is to write **proportions** so that units cancel.

18 pints = _____ gallons

$$\frac{\overset{9}{\cancel{18}}\ \cancel{pt}}{1} \cdot \frac{1\ \cancel{qt}}{\underset{1}{\cancel{2}}\ \cancel{pt}} \cdot \frac{1\text{ gal}}{4\ \cancel{qt}} = \frac{9}{4}\text{ gal} = 2\frac{1}{4}\text{ gal}$$

VOCABULARY

capacity
the amount of dry or liquid material a container can hold

customary system
a measurement system that includes units such as foot, pound, and quart

length
a measurement of the distance between two points

proportion
an equation of the form $\frac{a}{b} = \frac{c}{d}$ stating that two ratios are equivalent

weight
a measurement that tells how heavy or light an object is

Example 1

Convert.

90 in. = ______ ft

1. Are the units changing to a smaller unit or a larger unit? **larger**
2. Multiply or divide? **divide**
3. Write the equivalency needed.

   1 ft = 12 in.
4. Divide 90 by 12.

   90 ÷ 12 = 7.5

So, 90 in. = 7.5 ft.

YOUR TURN!

Convert.

3.5 mi = ______ yd

1. Are you changing to a smaller unit or a larger unit? ________
2. Multiply or divide? ________
3. Write the equivalency needed.

   ____
4. ____________ by ________.

   ____________ = ________

So, 3.5 mi = ________ yd.

Example 2

Convert.

3 pt = ______ fl oz

1. Write the equivalencies needed.

   1 pt = 2 c

   1 c = 8 fl oz
2. Set up a proportion so that units cancel.

   $$\frac{3\ \not{pt}}{1} \cdot \frac{2\ \not{c}}{1\ \not{pt}} \cdot \frac{8\text{ fl oz}}{1\ \not{c}}$$
3. Multiply and simplify. Convert any fractions to decimals.

   $$\frac{3}{1} \cdot \frac{2}{1} \cdot \frac{8\text{ fl oz}}{1} = 48\text{ fl oz}$$

So, 3 pt = 48 fl oz.

YOUR TURN!

Convert.

14 pt = ______ gal

1. Write the equivalencies needed.

   ______ qt = ______ pt

   ______ gal = ______ qt
2. Set up a proportion so that units cancel.

   $$\frac{14\text{ pt}}{1} \cdot \frac{\square}{\square} \cdot \frac{\square}{\square}$$
3. Multiply and simplify. Convert any fractions to decimals.

   ______ • ______ • ______ = ______

So, 14 pt = ________.

GO ON

Guided Practice

Convert each measurement.

1. 31.5 ft = ________ yd

   Changing to a ________ unit,

   so ________.

   Use the equivalency _____ yd = _____ ft.

   31.5 ________ = ________ yd

2. 6.5 qt = ________ pt

   Changing to a ________ unit,

   so ________.

   Use the equivalency _____ qt = _____ pt.

   6.5 ________ = ________ pt

3. 14 mi = ________ ft

   Changing to a ________ unit,

   so ________.

   Use the equivalency _____ = ________.

   ________ = ________

4. 128 oz = ________ c

   Changing to a ________ unit,

   so ________.

   Use the equivalency _____ = ________.

   ________ = ________

Step by Step Practice

5. Convert.

   360 fl oz = ________ qt

   Step 1 Write the equivalencies needed.

   1 c = ________ fl oz

   1 pt = ________ c

   1 qt = ________ pt

   Step 2 Set up a proportion so that units cancel.

   $$\frac{360 \text{ fl oz}}{1} \cdot \frac{\square}{\square} \cdot \frac{\square}{\square} \cdot \frac{\square}{\square}$$

   Step 3 Multiply and simplify.

   ________ = ________ = ________

Convert each measurement.

6 64,000 oz = ______ T

$$\frac{64{,}000\text{ oz}}{1} \cdot \frac{1\text{ lb}}{\square} \cdot \frac{1\text{ T}}{\square} =$$

$$\frac{\square}{\square} = \_\_\_\_$$

7 522 in. = ______ yd

$$\frac{522\text{ in.}}{1} \cdot \frac{1\text{ ft}}{\square\text{ in.}} \cdot \frac{\square}{\square\text{ ft}} =$$

$$\frac{\square}{\square} = \_\_\_\_\frac{\square}{\square}$$

Step by Step Problem-Solving Practice

Solve.

8 **COOKING** Mrs. Henderson needs 3 pints of chicken broth to make her vegetable soup for a party. The recipe makes soup for 8 people. Mrs. Henderson needs to make soup for 32 guests. Cans of chicken broth are sold in 1 quart cans. How many cans does she need to buy?

______ guests ÷ ______ people = ______ recipes

She needs to make ______ recipes.

______ pints • ______ recipes = ______ pints

She needs ______ pints of broth.

Mrs. Henderson needs to buy ______ of chicken broth.

Check off each step.

______ **Understand: I underlined key words.**

______ **Plan: To solve the problem, I will** ____________________.

______ **Solve: The answer is** ____________________.

______ **Check: I checked my answer by** ____________________.

GO ON

Skills, Concepts, and Problem Solving

Convert each measurement.

9. 15 c = ________ pt

10. 8 qt = ________ c

11. 5.5 mi = ________ yd

12. 3,000 lb = ________ T

13. 2 T = ________ oz

14. 72 in. = ________ yd

15. 2.5 gal = ________ pt

16. 124 in. = ________ ft

Solve.

17. **ZOO** Shanti, the new elephant that arrived at the city zoo, weighs 5.6 tons. How many pounds does Shanti weigh?

____

18. **COOKING** Henry needed 10 pints of juice to make fruit punch for a party. How many gallons of juice did he need?

____

19. **TRAVEL** Miranda walks 7,920 feet to school every morning. How far does Miranda walk in miles?

____

Vocabulary Check **Write the vocabulary word that completes each sentence.**

20. The ________________ includes units such as foot, pound, and quart.

21. ________________ is the amount of dry or liquid material a container can hold.

22. **Reflect** Why do you multiply when converting from a larger unit to a smaller unit?

____

____

STOP

Lesson 5-2

The Metric System

KEY Concept

The **metric system** is based on powers of 10. The **meter** is the base unit of length.

Unit of Length	Abbreviation	Number of Meters
millimeter	mm	0.001 m
centimeter	cm	0.01 m
decimeter	dm	0.1 m
meter	m	1 m
dekameter	dkm	10 m
hectometer	hm	100 m
kilometer	km	1,000 m

The **liter** is the base unit of **capacity**.

Unit of Capacity	Abbreviation	Number of Liters
milliliter	mL	0.001 L
centiliter	cL	0.01 L
deciliter	dL	0.1 L
liter	L	1 L
dekaliter	dkL	10 L
hectoliter	hL	100 L
kiloliter	kL	1,000 L

The **gram** is the base unit of **mass**.

Unit of Mass	Abbreviation	Number of Grams
milligram	mg	0.001 g
centigram	cg	0.01 g
decigram	dg	0.1 g
gram	g	1 g
dekagram	dkg	10 g
hectogram	hg	100 g
kilogram	kg	1,000 g

The metric system uses prefixes to indicate that measurement units are larger or smaller than the base unit.

VOCABULARY

capacity
the amount of dry or liquid material a container can hold

gram
a base unit of measurement for mass in the metric system

liter
a base unit of measurement for capacity in the metric system

mass
the amount of matter in an object

meter
a base unit of measurement for length in the metric system

metric system
a measurement system that includes units such as meter, gram, and liter

GO ON

Example 1

Convert.

8 kL = ________ L

1. Are the units changing to a smaller unit or a larger unit? **smaller**
2. Multiply or divide? **multiply**
3. Write the equivalency needed.

   1 kL = 1,000 L

4. Multiply 8 by 1,000.

   8 • 1,000 = 8,000

So, 8 kL = 8,000 L.

YOUR TURN!

Convert.

240 cg = ________ g

1. Are the units changing to a smaller unit or a larger unit? ________
2. Multiply or divide? ________
3. Write the equivalency needed.

   ________ = ________.

4. ________ by ________.

   ____

So, 240 cg = ________ g.

Example 2

Convert.

140 km = ________ cm

1. Are the units changing to a smaller unit or a larger unit? **smaller**
2. Do you multiply or divide? **multiply**
3. Use a model.

÷ 1,000 ÷ 100 ÷ 10

km m cm mm

× 1,000 × 100 × 10

140 • 1,000 • 100 = 14,000,000

So, 140 km = 14,000,000 cm.

YOUR TURN!

Convert.

4,806 mL = ________ L

1. Are the units changing to a smaller unit or a larger unit? ________
2. Multiply or divide? ________
3. Use a model.

÷ 1,000 ÷ 100 ÷ 10

KL L cL mL

× 1,000 × 100 × 10

4,806 ________ = ________

So, 4,806 mL = ________ L.

Guided Practice

Convert each measurement.

1 96 m = ________ km

Changing to a ________ unit,

so ________.

Use the equivalency

________ m = ________ km.

96 ____________ = ________ km

2 15 m = ________ cm

Changing to a ________ unit,

so ________.

Use the equivalency

________ cm = ________ m.

15 ____________ = ________ cm

3 230 mL = ________ L

Changing from ________ to ________

units, so ________.

Use the equivalency

________ = ________.

____________ = ________ L

4 70 kL = ________ L

Changing from ________ to ________

units, so ________.

Use the equivalency

________ = ________.

____________ = ________ L

Step by Step Practice

5 Convert. 25 kg = ________ cg

Step 1 Are the units changing to a smaller unit or a larger unit?

____

Step 2 Do you multiply or divide? ________

Step 3 Use the model.

÷ 1,000 ÷ 100 ÷ 10

kg g cg mg

1,000 × 100 × 10 ×

25 ____________ = ____________.

Step 4 25 kg = ____________ g

GO ON

Convert each measurement.

6 15 mL = __________ kL

÷ 1,000 ÷ 100 ÷ 10

kL L cL mL

× 1,000 × 100 × 10

Change from __________ to __________ units.

__________ by _____, _____ and then by __________.

15 ____________________ = ____________________

7 67.4 km = __________ cm

÷ 1,000 ÷ 100 ÷ 10

km m cm mm

× 1,000 × 100 × 10

Change from __________ to __________ units.

__________ by __________ and then by __________.

____________________ = ____________________

Step by Step Problem-Solving Practice

Solve.

8 **ARCHITECTURE** The Sears Tower, one of the world's tallest buildings, measures 0.519 km. The Buri Dabai, which stands 0.818 km, is another one of the world's tallest buildings. What is the difference in the buildings' heights expressed in meters?

The difference in the heights is _______ km − _______ km = _______ km.

Convert from km to m:

____________ = _______

Check off each step.

_____ **Understand: I underlined key words.**

_____ **Plan: To solve the problem, I will** ____________________.

_____ **Solve: The answer is** ____________________.

_____ **Check: I checked my answer by** ____________________.

Skills, Concepts, and Problem Solving

Convert each measurement.

9 13 g = ______________ kg

10 9 m = ______________ mm

11 2,000 mL = ______________ kL

12 26 kg = ______________ mg

13 850 cm = ______________ km

14 6 L = ______________ mL

Solve.

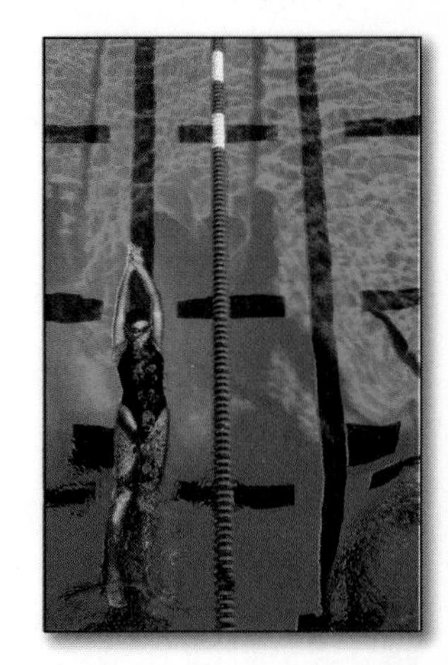

15 **SWIMMING** The length of an Olympic-size swimming pool is 50 meters. If a race is 4 lengths of the pool, how many kilometers is the race?

____

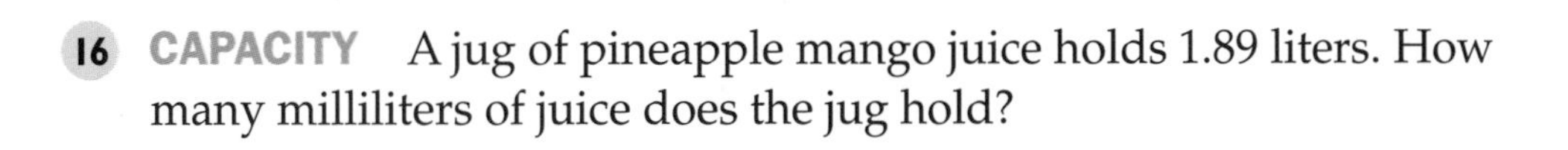

16 **CAPACITY** A jug of pineapple mango juice holds 1.89 liters. How many milliliters of juice does the jug hold?

____

17 **MASS** The average weight for a gorilla is 350 kilograms. What is the average weight for a gorilla in milligrams?

____

Vocabulary Check **Write the vocabulary word that completes each sentence.**

18 The______________________ includes units such as meter, kilogram, and liter.

19 The base unit of measurement for mass in the metric system is

a(n) ______________________.

20 **Reflect** Describe the movement of the decimal point when you convert in the metric system. Explain how you know the direction that the decimal point moves.

____

____

____

____

Chapter 5

Progress Check 1 (Lessons 5-1 and 5-2)

Convert each measurement.

1. 12 c = ______________ qt
2. 2 qt = ______________ oz
3. 1.4 mi = ______________ yd
4. 5,000 oz = ______________ T
5. 92 oz = ______________ lb
6. 126 in. = ______________ yd
7. 251 g = ______________ kg
8. 9 m = ______________ mm
9. 1.32 kL = ______________ mL
10. 91.3 kg = ______________ mg
11. 444.3 m = ______________ km
12. 12 L = ______________ mL

Solve.

13. **EXERCISE** Danny ran 8 times around a quarter-mile track. What is the total number of feet that he ran?

____

14. **COMMUNITY** Alisha has 2 liters of bath oil. She wants to divide it equally between her four friends. How many milliliters will each friend receive?

____

15. **TRAVEL** Julian is driving in Canada. The speed limit is 100 kilometers per hour. If he drives the speed limit, how many meters will he travel in two hours?

____

16. **ANIMALS** Talia has a horse that weighs $\frac{2}{3}$ of a ton. How many pounds does the horse weigh?

____

Lesson 5-3

Convert Between Systems

KEY Concept

Convert between the **customary** and **metric** system using the information below.

Customary Unit	Approximate Metric Equivalent
1 in.	2.54 cm
1 ft	30.48 cm or 0.3048 m
1 yd	0.914 m
1 mi	1.609 km
1 oz	28.350 g
1 lb	454 g or 0.454 kg
1 fl oz	29.574 mL
1 qt	0.946 L
1 gal	3.785 L

To convert from the customary system to the metric system, multiply by the conversion factors from the table.

To convert to the metric system to the customary system, divide by the conversion factors from the table.

VOCABULARY

customary system
a measurement system that includes units such as foot, pound, and quart

metric system
a measurement system that includes units such as meter, gram, and liter

The symbol ≈ means the answer is an estimate, and therefore not an exact equivalency. Round conversions to the nearest thousandth.

Example 1

Convert. Round to the nearest thousandth.

20 gal ≈ ________ L

1. Use the conversion factor.

   1 gal ≈ 3.785 L

2. Multiply.

   20 • 3.785 ≈ 75.7

So, 20 gal ≈ 75.7 L.

YOUR TURN!

Convert. Round to the nearest thousandth.

180 lb ≈ ________ g

1. Use the conversion factor.

   ________ ≈ ________

2. Multiply.

   ________ • ________ ≈ ________

So, 180 lb ≈ ________ g.

GO ON

Example 2

Convert. Round to the nearest thousandth.

6 kg ≈ ________ lb

1. Use the conversion factor.

   1 lb ≈ 0.454 kg

2. This factor is in the reverse order of the conversion factor, so divide.

   6 ÷ 0.454 ≈ 13.216

So, 6 kg ≈ 13.216 lb.

YOUR TURN!

Convert. Round to the nearest thousandth.

23 g ≈ ________ oz

1. Use the conversion factor.

   ________ ≈ ________

2. This factor is in the reverse order of the conversion factor, so ________.

   ________ ≈ ________

So, 23 g ≈ ________ oz.

Guided Practice

Convert. Round to the nearest thousandth.

1. 8 yd ≈ ________ m

   Use ____ yd ≈ ________ m to multiply.

   ________ ≈ ________

2. 5 oz ≈ ________ g

   Use ____ oz ≈ ________ g to multiply.

   ________ ≈ ________

3. 6.5 ft ≈ ________ cm

   Use ____ ft ≈ ________ cm to ________.

   ________ ≈ ________

4. 14.5 lb ≈ ________ kg

   Use ____ lb ≈ ________ kg to ________.

   ________ ≈ ________

Step by Step Practice

5. Convert. Round to the nearest thousandth. 3,800 L ≈ ________ gal

   Step 1 Use the conversion factor. ________ ≈ ________

   Step 2 This factor is in the reverse order of the conversion factor,

   so ________.

   ________ ≈ ________

   Step 3 3,800 L ≈ ________ gal

Convert. Round to the nearest thousandth.

6. 390 cm ≈ __________ in.

   Use _____ in. ≈ __________ cm to divide.

   ______________ ≈ __________

7. 57 kg ≈ __________ lb

   Use _____ lb ≈ __________ kg to divide.

   ______________ ≈ __________

8. 204 L ≈ ______________ qt

   Use _____ qt ≈ ________ L to ________.

   ______________ ≈ __________

9. 8.4 m ≈ ______________ yd

   Use _____ m ≈ ________ yd to ________.

   ______________ ≈ __________

Step by Step Problem-Solving Practice

Solve.

10. **AMUSEMENT PARK** The Son of Beast, a roller coaster at Kings Island in Ohio, is the longest wooden roller coaster in the United States. It is 7,032 feet long. How long is the Son of Beast measured in meters? Round the answer to the nearest thousandths.

    Use the conversion factor __________ ft ≈ ______________ m.

    To solve this problem, I will _________________.

    ________________ ≈ _________________

    ________________ ft ≈ _________________ m

    Check off each step.

    ______ **Understand: I underlined key words.**

    ______ **Plan: To solve the problem, I will** ____.

    ______ **Solve: The answer is** ____.

    ______ **Check: I checked my answer by** ____.

GO ON

Skills, Concepts, and Problem Solving

Convert. Round to the nearest thousandth.

11 72 m ≈ ____________ yd

12 88 mL ≈ ____________ fl oz

13 19 in. ≈ ____________ cm

14 42 gal ≈ ____________ L

15 45,000 L ≈ ____________ gal

16 350 g ≈ ____________ lb

17 30 lb ≈ ____________ kg

18 7.5 mi ≈ ____________ km

Solve. Round to the nearest thousandth.

19 **EXERCISE** Every afternoon Terrell runs 2.5 miles. How many kilometers does he run each day?

____

20 **ANIMALS** Giraffes have long tongues which can be extended more than 45 cm to use for grasping food. How many inches can a giraffe extend its tongue?

____

Vocabulary Check **Write the vocabulary word that completes each sentence.**

21 The ____________ system includes units such as feet, pounds, and quarts.

22 The ____________ system includes units such as meters, grams, and liters.

23 **Reflect** How can you tell if the two units being compared are close in size? Give an example.

____

____

____

STOP

Lesson 5-4

Perimeter

KEY Concept

One way to find the **perimeter** of a figure is to use graph paper.

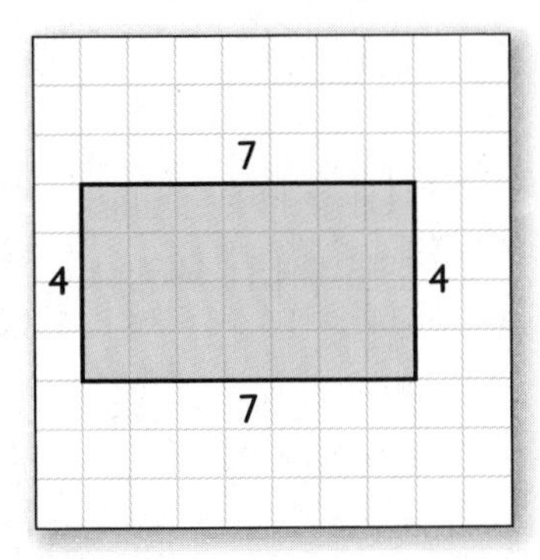

P = side + side + side + side
$P = 4 + 7 + 4 + 7 = 22$

So, the perimeter is 22 units.

To find the perimeter of a figure without graph paper, add the lengths of its sides.

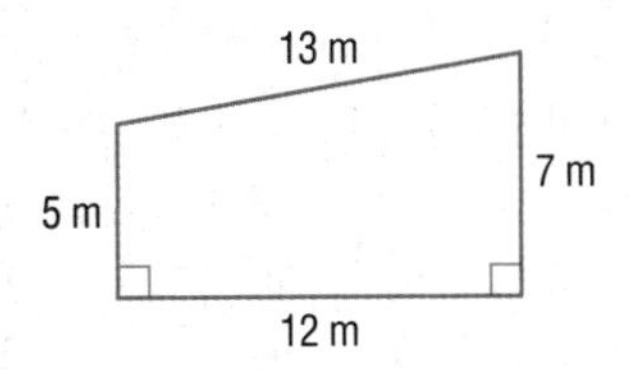

$P = 5 + 12 + 7 + 13 = 37$

So, the perimeter is 37 m.

To find the perimeter of a **regular polygon**, you can multiply one length of one side by the number of sides.

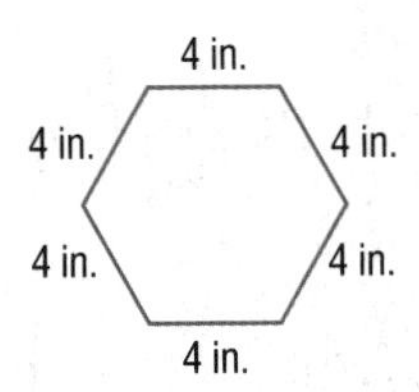

P = 4 in. • 6 sides = 24 in.

So, the perimeter is 24 in.

VOCABULARY

perimeter
the distance around a shape or a figure

polygon
a closed plane figure

regular polygon
a polygon with all sides equal in length

Perimeter is measured in linear units. Linear units measure lengths such as centimeters, meters, inches, yards, and feet.

GO ON

Example 1

Find the perimeter.

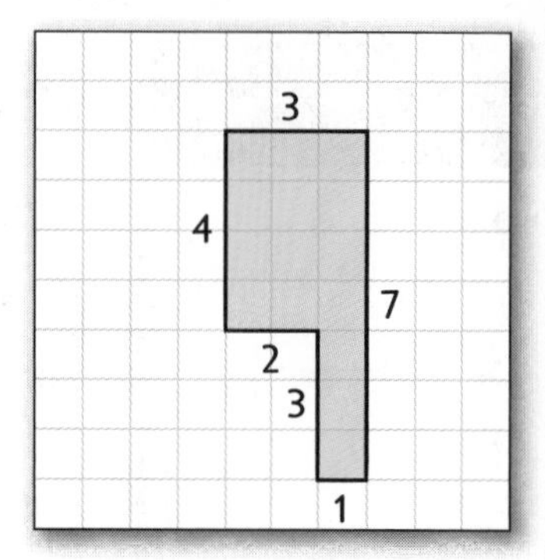

1. The figure has 6 sides. Count the length of each side.
2. Add the side lengths.

   3 + 7 + 1 + 3 + 2 + 4 = 20

3. The perimeter is 20 units.

YOUR TURN!

Find the perimeter.

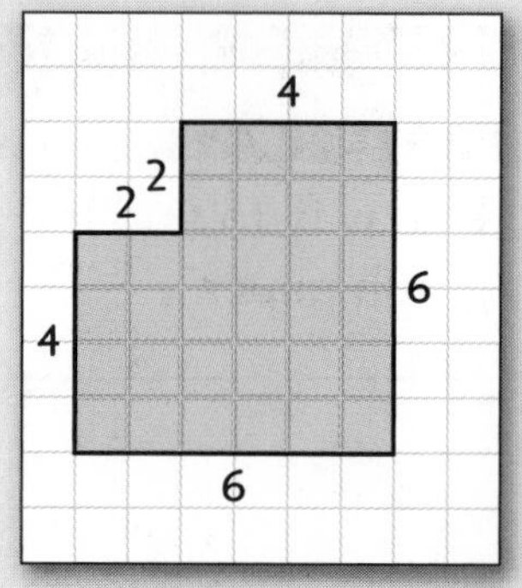

1. The figure has _____ sides. Count the length of each side.
2. Add the side lengths.

   _______________________ = _____

3. The perimeter is _____ units.

Example 2

Find the perimeter.

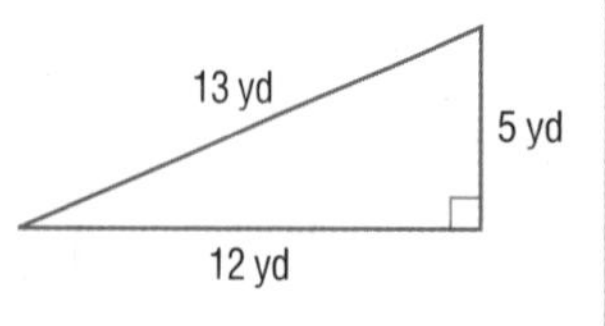

1. The figure has 3 sides.
2. Add the side lengths.

   13 + 12 + 5 = 30

3. The perimeter is 30 yd.

YOUR TURN!

Find the perimeter.

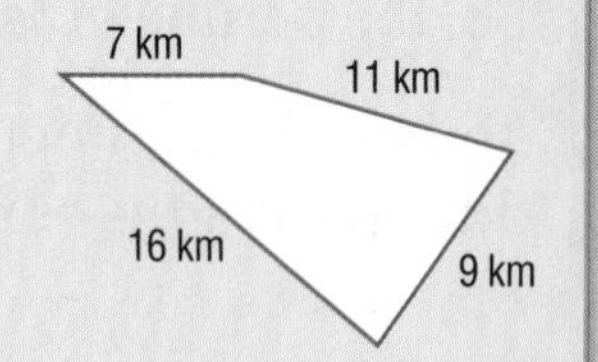

1. The figure has _____ sides.
2. Add the side lengths.

   _______________________ = _____

3. The perimeter is _____ km.

Example 3

Find the perimeter.

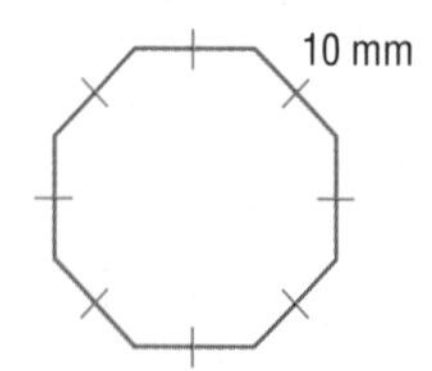

1. The figure is a regular polygon with 8 sides.
2. The length of each side is 10 mm.

The single tick marks mean the sides are the same length.

3. Multiply.

   8 • 10 = 80

4. The perimeter is 80 mm.

YOUR TURN!

Find the perimeter.

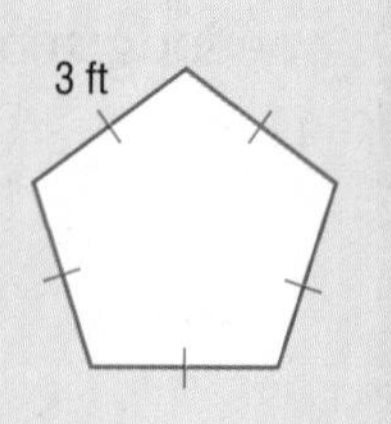

1. The figure is a __________ polygon with _____ sides.

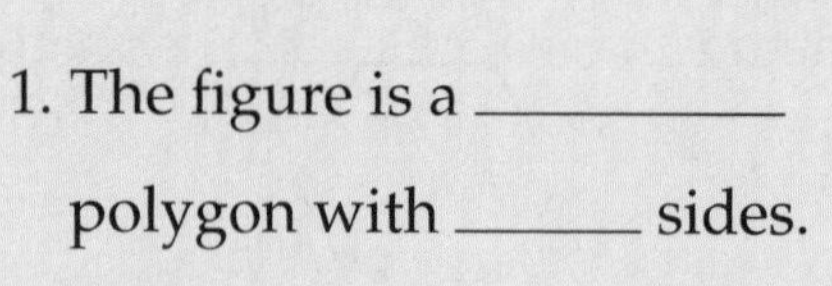

2. The length of each side is _____.
3. Multiply.

   _____ • _____ = _____

4. The perimeter is _____ ft.

Guided Practice

Find the perimeter of each figure.

1

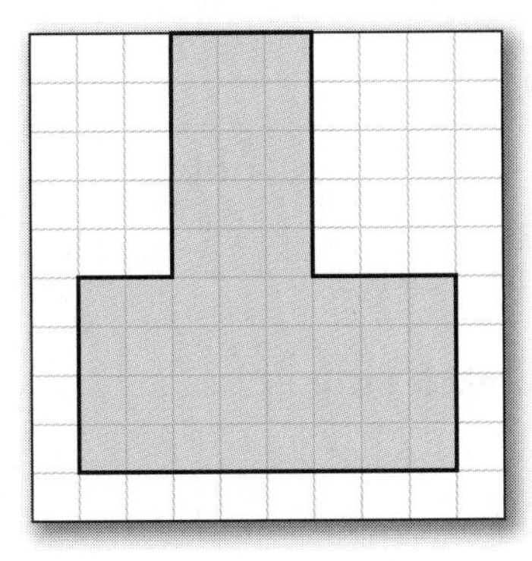

The figure has ______ sides.

______________________ = ______

The perimeter is ______ units.

2

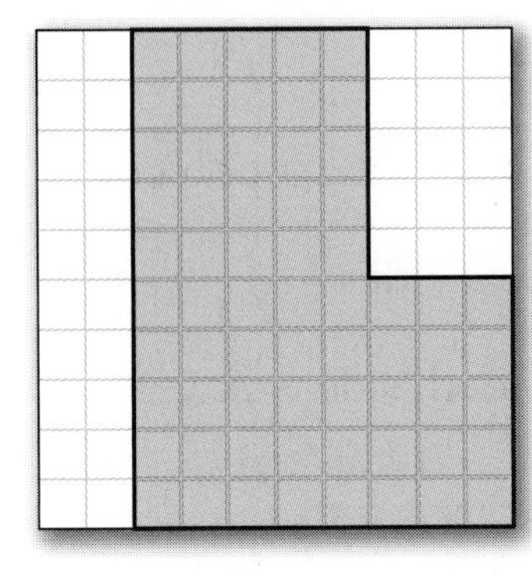

The figure has ______ sides.

______________________ = ______

The perimeter is ______ units.

3

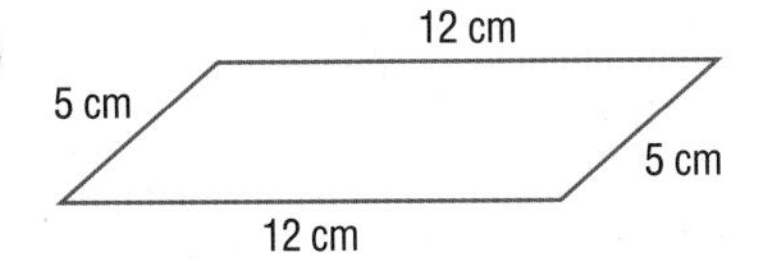

The figure has ______ sides.

______________________ = ______

The perimeter is ______ cm.

4

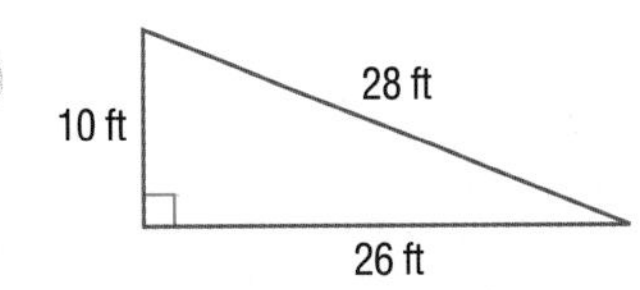

The figure has ______ sides.

______________________ = ______

The perimeter is ______ ft.

Step by Step Practice

5 Find the perimeter.

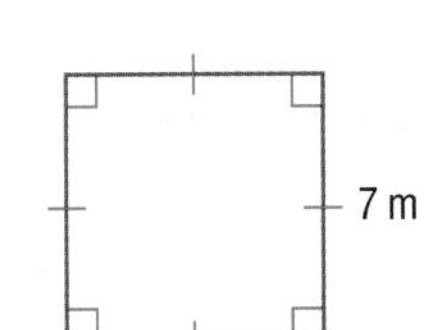

Step 1 The figure is a ________ polygon with ________ sides.

Step 2 The length of each side is ________.

Step 3 Multiply.

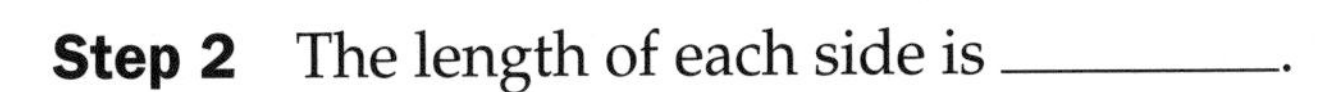

________ • ________ = ________

Step 4 The perimeter is ________ m.

GO ON

Find the perimeter of each figure.

6 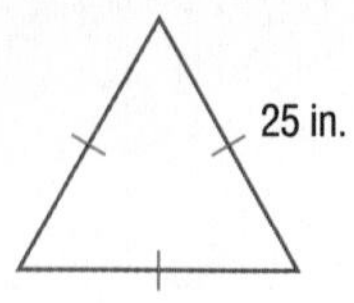

The figure is a ________ polygon.

It has ________ sides.

The length of each side is ________.

________ = ________

The perimeter is ________ in.

7 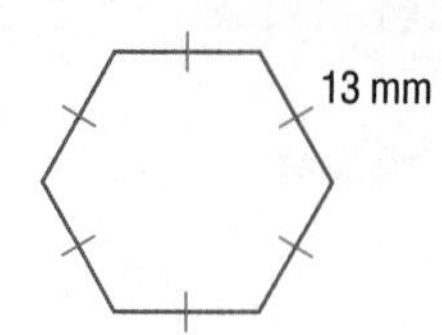

The figure is a ________ polygon.

It has ________ sides.

The length of each side is ________.

________ = ________

The perimeter is ________ mm.

Step by Step Problem-Solving Practice

Solve.

8 **DECORATING** Megan has a regular octagonal mirror that she is going to glue wooden trim around. If each side of the mirror is 10 inches, how much trim will she need?

An octagon has ______ sides.

______ • ______ = ______ inches of wood trim

Check off each step.

______ **Understand: I underlined key words.**

______ **Plan: To solve the problem, I will** ________________________________.

______ **Solve: The answer is** ________________________________.

______ **Check: I checked my answer by** ________________________________.

Skills, Concepts, and Problem Solving

Find the perimeter of each figure.

9

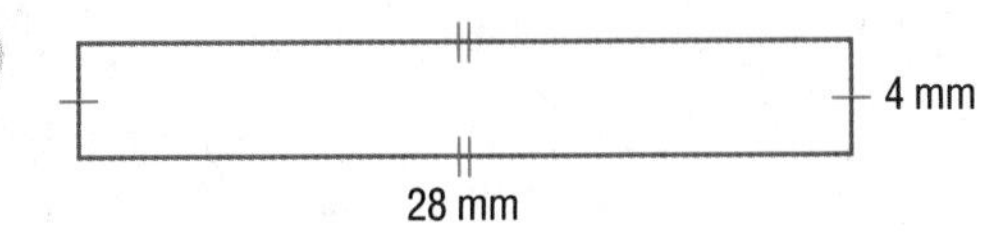

$P =$ ____________

10

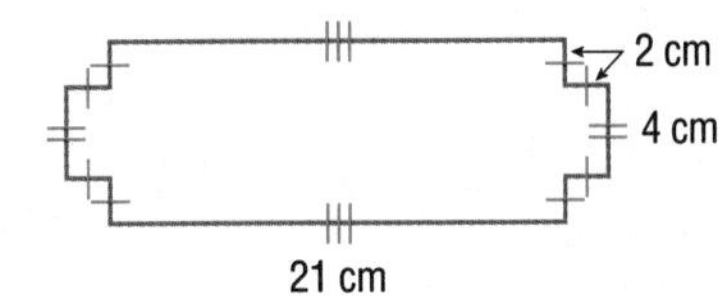

$P =$ ____________

11

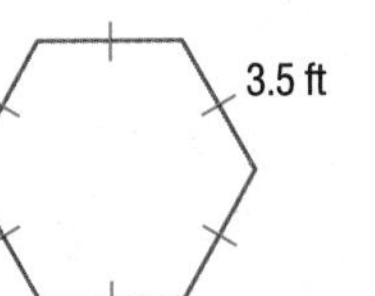

$P =$ ____________

12

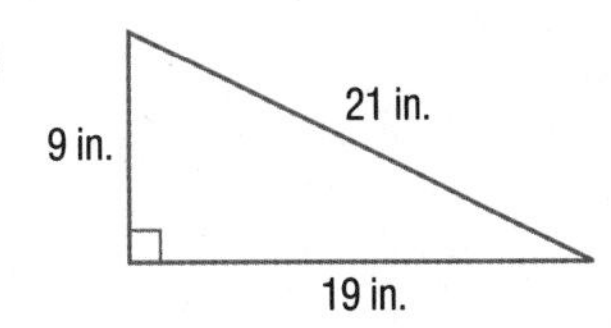

$P =$ ____________

Solve.

13 **SPORTS** The dimensions of a rectangular field for men's lacrosse are 110 yards long and 60 yards wide. What is the perimeter of the field?

____

14 **ART** Mr. Patel wants to paint a mosaic trim around the perimeter of his hot tub. It is shaped like a regular pentagon. If one side length of his hot tub is 1.2 meters, what is the perimeter?

____

15 **FENCING** The Larson family has decided to install an invisible fence for their dog. Using the diagram, how many feet of fencing will they need if they install the fence along the red borders?

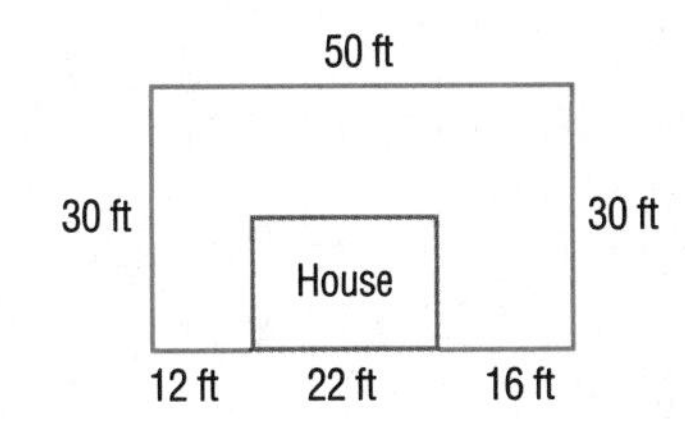

____

Vocabulary Check **Write the vocabulary word that completes each sentence.**

16 ____________ is the distance around a shape or a figure.

17 A(n) ____________ has sides of equal length.

18 **Reflect** Give the dimensions of four different rectangles that have a perimeter of 36 units.

____

____

STOP

Chapter 5

Progress Check 2 (Lessons 5-3 and 5-4)

Convert. Round to the nearest thousandth.

1. 12 in. = ____________ cm
2. 18.925 L = ____________ gal
3. 8 fl oz ≈ ____________ mL
4. 9 km = ____________ mi
5. 1 T ≈ ____________ kg
6. 900 g ≈ ____________ oz
7. 200 m = ____________ ft
8. 0.5 L ≈ ____________ fl oz
9. 12 oz = ____________ g
10. 2,000 cm = ____________ ft
11. 150 L ≈ ____________ gal
12. 400 lb ≈ ____________ kg

Find the perimeter of each figure.

13. 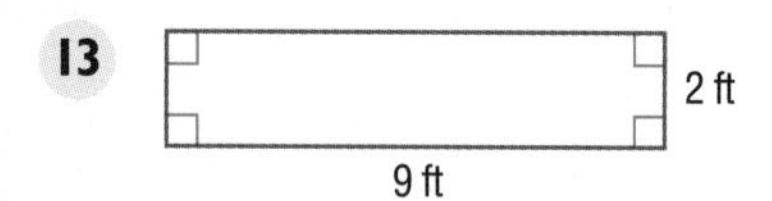

$P =$ ____________

14. 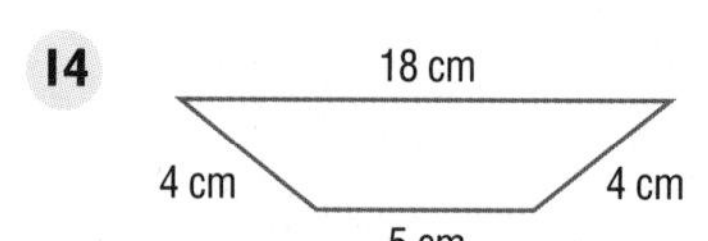

$P =$ ____________

15. 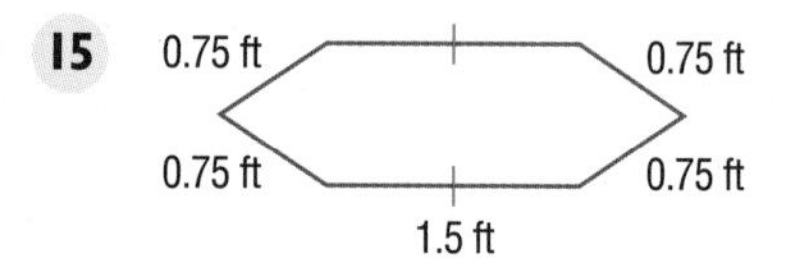

$P =$ ____________

16. 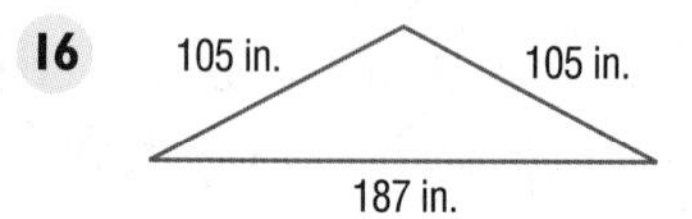

$P =$ ____________

Solve. Round to the nearest thousandth.

17. **AQUARIUM** Akio purchased a 250-liter aquarium. About how many gallons of water does it hold?

____

18. **PETS** Michelle feeds her hamster 14 grams of food a day. About how many ounces of food does it eat per day?

____

19. **HEALTH** Instructions for a cold medicine state that you should take 1 fluid ounce of medicine three times a day. About how many milliliters of medicine will be taken each day?

____

Lesson 5-5

Area

KEY Concept

The **area** of a figure is how much of a region or surface it covers. Area is measured in square units, such as in^2, cm^2, or $units^2$.

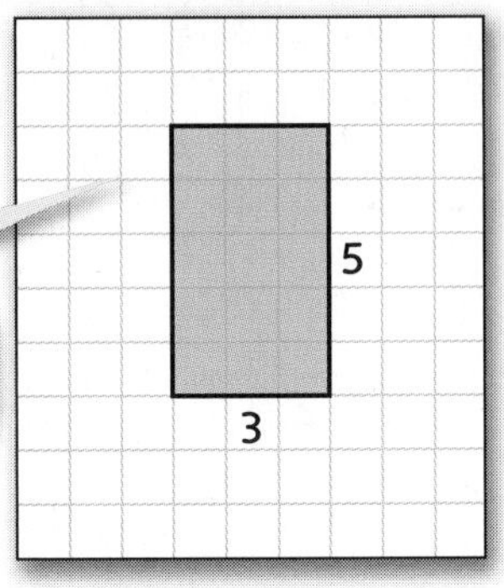

The rectangle covers 15 blocks.

You can find the area of a **rectangle** without graph paper by multiplying the length times the width.

$$\text{Area} = \ell \cdot w \qquad \text{Area} = 3 \cdot 5 = 15$$

The area of the rectangle is 15 square units or 15 $units^2$.

In a **parallelogram**, the base (b) is the same as the length of a rectangle. The height (h) is the measure of a line that is perpendicular to the base.

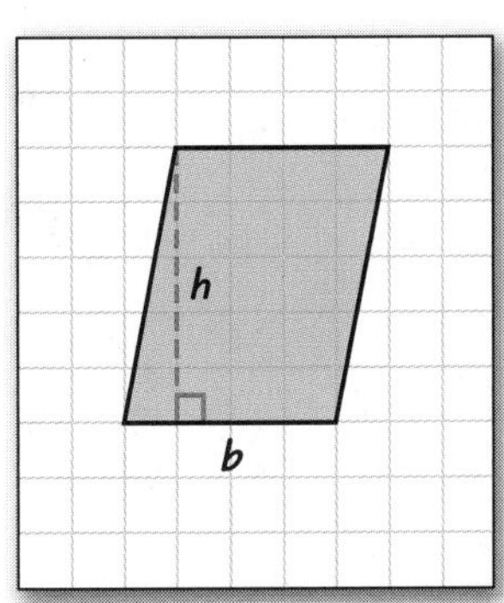

You can also find the area of a parallelogram by multiplying the base times the height.

$$\text{Area} = b \cdot h \qquad \text{Area} = 4 \cdot 5 = 20$$

The area of the parallelogram is 20 square units or 20 $units^2$.

The area of a **triangle** is one-half the area of a parallelogram. To find the area of the triangle, divide the area of a parallelogram in half.

If the triangle has a right angle, the area is $\frac{1}{2}$ the area of a rectangle.

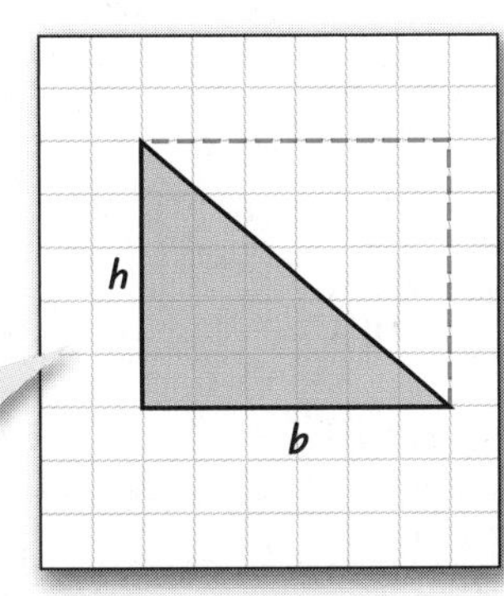

$$\text{Area} = \frac{1}{2}bh \qquad \text{Area} = \frac{1}{2}(6)(5) = 15$$

The area of the triangle is 15 square units or 15 $units^2$.

VOCABULARY

area
the number of square units needed to cover a surface enclosed by a geometric figure

parallelogram
a quadrilateral with opposite sides parallel. Any side of a parallelogram may be called the base

rectangle
a quadrilateral with four right angles

triangle
a polygon with three sides and three angles

GO ON

Example 1

Find the area.

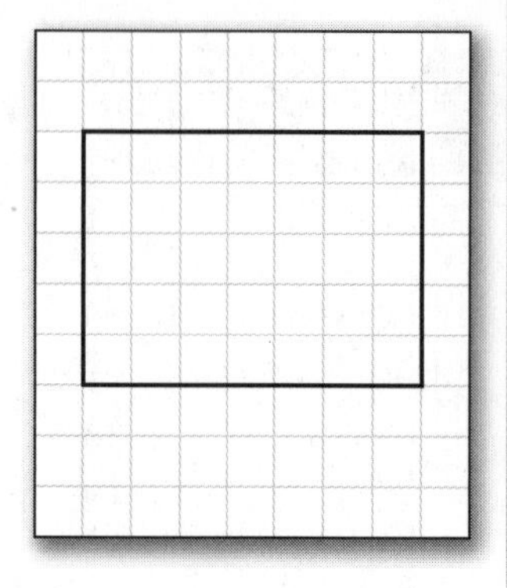

1. Find the length. 7

2. Find the width. 5

3. Multiply the length by the width.

   Area $= \ell \cdot w$

   $= 7 \cdot 5 = 35$

4. The area is 35 units². Count the units to check your answer.

YOUR TURN!

Find the area.

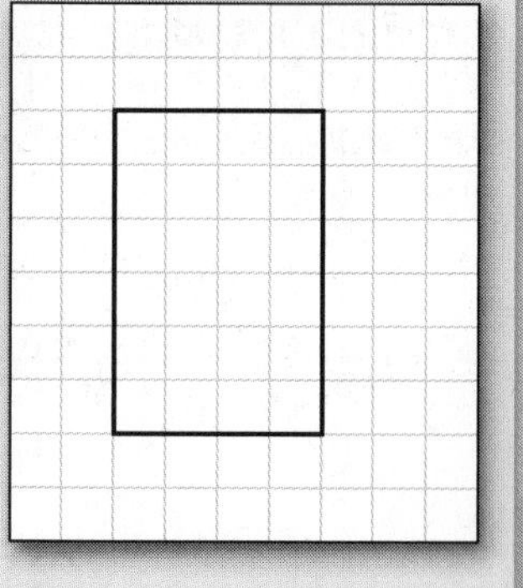

1. Find the length.

   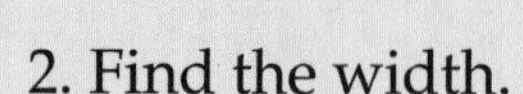

   ______

2. Find the width.

   ______

3. Multiply the length by the width.

   Area $= \ell \cdot w$

   $=$ ______ $\cdot$ ______ $=$ ______

4. The area is __________. Count the units to check your answer.

Example 2

Find the area.

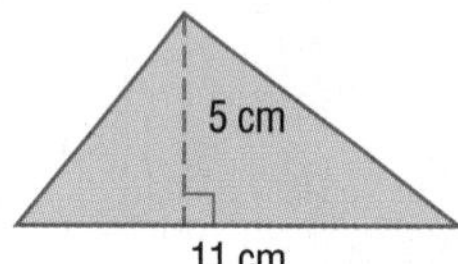

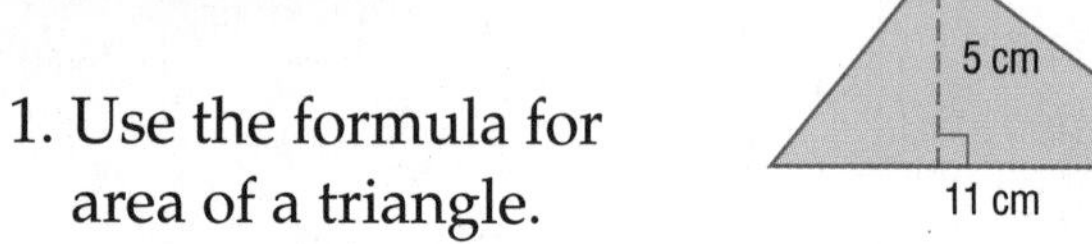

1. Use the formula for area of a triangle.

   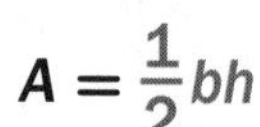

   $A = \frac{1}{2}bh$

2. Substitute 11 for b and 5 for h.

   $\frac{1}{2}(11)(5) = \frac{1}{2}(55) = 27.5$

3. The area is 27.5 cm².

YOUR TURN!

Find the area.

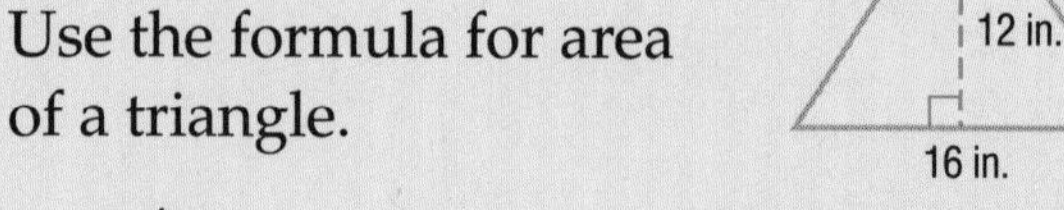

1. Use the formula for area of a triangle.

   $A =$ ______

2. Substitute ______ for b and ______ for h.

   ____

3. The area is ________.

Guided Practice

Find the area.

1 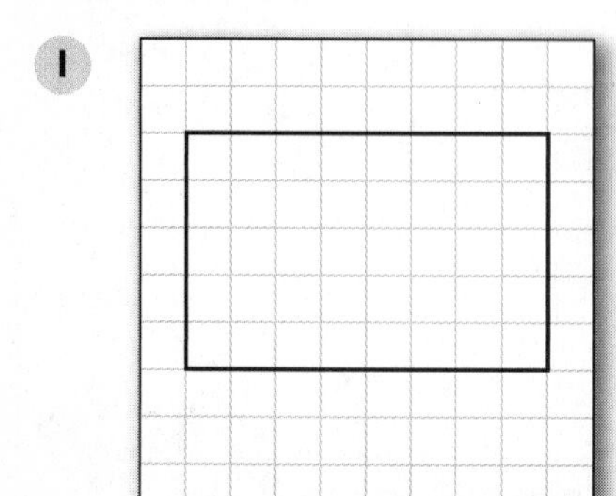

length = ______ units

width = ______ units

$A = \ell \cdot w$

$=$ ______ $\cdot$ ______ $=$ ______

The area is __________.

Step by Step Practice

2 Find the area of the triangle shown at the right.

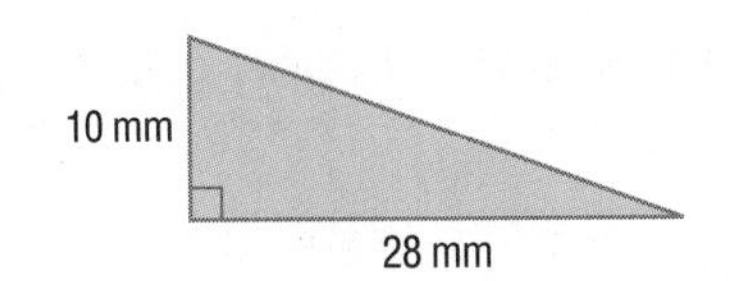

Step 1 Use the formula for the area of a triangle.

$A =$ ________

Step 2 Substitute ________ for b and ________ for h.

________ • ________ • ________ = ________

Step 3 Area = ________

Find the area of each figure.

3 parallelogram

base = 3.5 yd, height = 7.5 yd

$A =$ ________

$A =$ ________ • ________ = ________

4 parallelogram

base = 14 ft, height = 8.5 ft

$A =$ ________

$A =$ ________ • ________ = ________

Step by Step Problem-Solving Practice

Solve.

5 **FLOORING** Mr. Jensen is laying ceramic tile on the floor of his basement. His rectangular basement has a length of 30 feet and a width of 26 feet. Find the area of the floor of his basement.

The length is ______ feet. The width is ______ feet.

$A =$ ______ • ______ $A =$ ______ • ______

$A =$ ______ ft^2

Check off each step.

______ **Understand: I underlined key words.**

______ **Plan: To solve the problem, I will** ____________________.

______ **Solve: The answer is** ____________________.

______ **Check: I checked my answer by** ____________________.

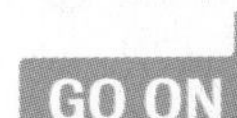

Skills, Concepts, and Problem Solving

Find the area of each figure.

6

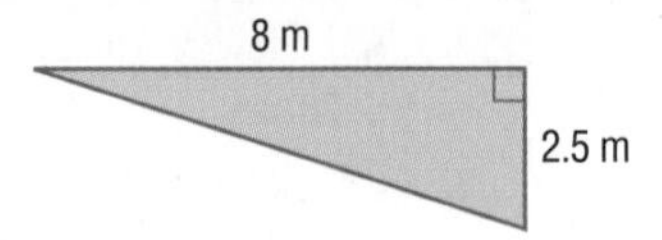

$A =$ ____________

7

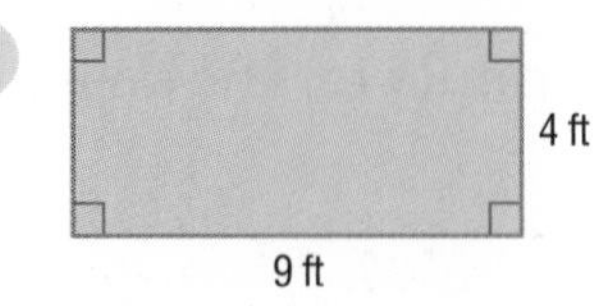

$A =$ ____________

8

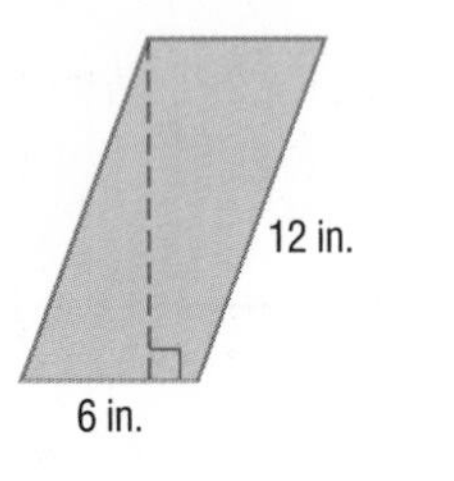

$A =$ ____________

9

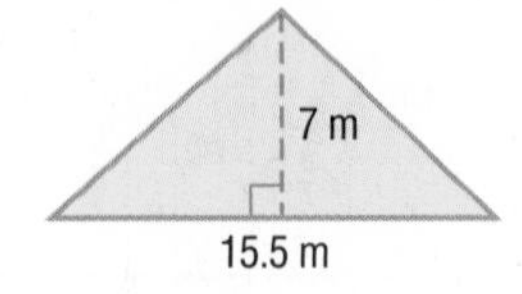

$A =$ ____________

10 parallelogram

$b = 22$ mm, $h = 13.5$ mm

$A =$ ____________

11 rectangle

$\ell = 13.25$ yd, $w = 10$ yd

$A =$ ____________

Solve. Write the answer in simplest form.

12 **LAND** An acre is a rectangular parcel of land that measures 660 feet by 66 feet. What is the area of an acre in square feet?

____

13 **GARDENING** Jolene is creating a new triangular garden on the side of her house. The base of her garden is 56 inches and the height is 48 inches. What is the area of her garden?

____

Vocabulary Check **Write the vocabulary word that completes each sentence.**

14 ____________ is the number of square units needed to cover a region or plane figure.

15 A(n) ____________ is a quadrilateral with opposite sides parallel.

16 **Reflect** Give the lengths and widths of four rectangles that have areas of 36 square units each.

____

STOP

Lesson 5-6

Volume

KEY Concept

Volume (V) is the measure of the space occupied by a solid.

The volume of a **prism** is the **area** of the **base** (B) times the height (h).

$$V = Bh$$

To find the volume of a **rectangular prism**, find the base which is the area of a rectangle.

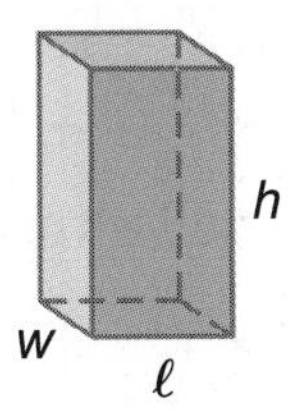

$$V = Bh$$
$$= (\ell w)h$$

To find the volume of a **triangular prism**, find the base which is the area of a triangle.

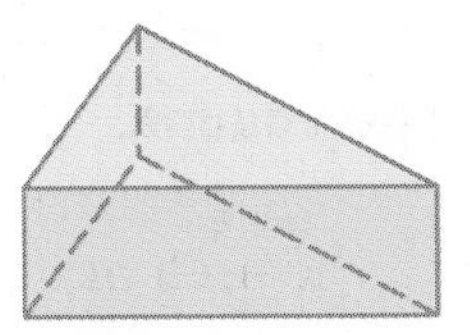

$$V = Bh$$
$$= \left(\frac{1}{2}Bh\right)h$$

The volume of a prism is measured in cubic units, such as in^3, cm^3, and ft^3.

VOCABULARY

area
the number of square units needed to cover a surface enclosed by a geometric figure

base
the faces on the top and bottom of a prism

cubic unit
a unit for measuring volume

prism
a polyhedron with two parallel, congruent faces called bases

rectangular prism
a three-dimensional figure with six faces that are rectangles

triangular prism
a prism whose bases are triangular with rectangles for sides

volume
the measure of space occupied by a solid region

GO ON

Example 1

Find the volume.

1. Find the area of the base, B.

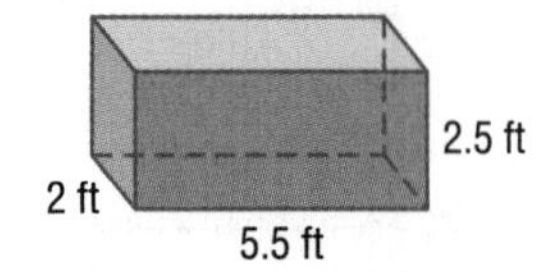

$B = \ell \cdot w$
$B = 5.5 \cdot 2$
$= 11 \text{ ft}^2$

2. Find the volume of the prism, V.

$V = B \cdot h$
$V = 11 \cdot 2.5$
$= 27.5 \text{ ft}^3$

YOUR TURN!

Find the volume.

1. Find the area of the base, B.

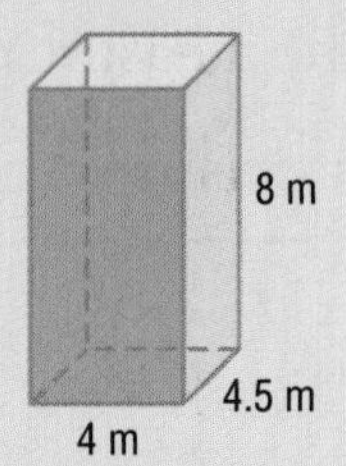

$B =$ ________ $\cdot$ ________

$B =$ ________ $\cdot$ ________

$=$ ________

2. Find the volume of the prism, V.

$V = B \cdot h$

$V =$ ________ $\cdot$ ________

$=$ ________

Example 2

Find the volume.

1. Find the area of the base, B.

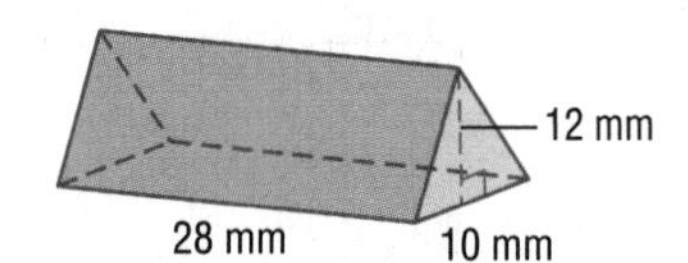

$B = \frac{1}{2} bh$

$B = \frac{1}{2}(10 \cdot 12)$

$= 60 \text{ mm}^2$

2. Find the volume.

$V = B \cdot h$
$V = 60 \cdot 28$
$= 1{,}680 \text{ mm}^3$

YOUR TURN!

Find the volume.

1. Find the area of the base, B.

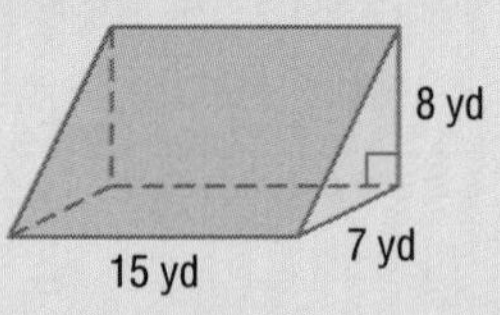

$B =$ ________

$B =$ ________ (________ $\cdot$ ________)

$=$ ________

2. Find the volume.

$V = B \cdot h$

$V =$ ________ $\cdot$ ________

$=$ ________

Guided Practice

Find the volume of each figure.

1

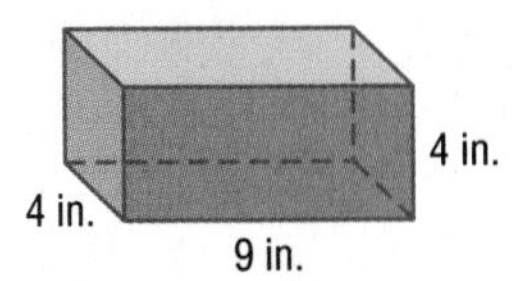

B = ________ • ________

B = ________ • ________ = ________

$V = B \cdot h$

V = ________ • ________ = ________

The volume of the prism is ________.

2

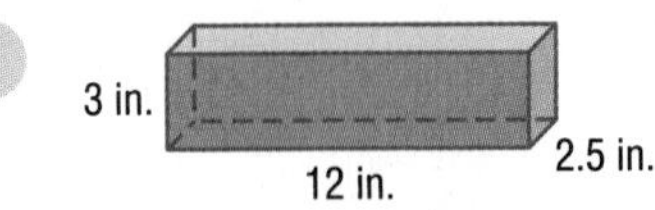

B = ________ • ________

B = ________ • ________ = ________

$V = B \cdot h$

V = ________ • ________ = ________

The volume of the prism is ________.

Step by Step Practice

3 Find the volume.

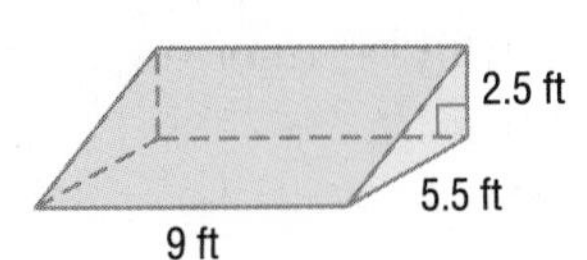

Step 1 Find the area of the base, B.

B = ________

B = ________ • ________ • ________

= ________

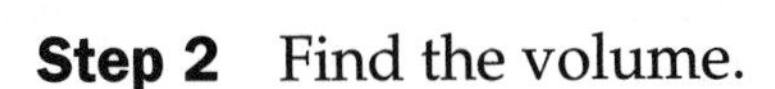

Step 2 Find the volume.

$V = B \cdot h$

V = ________ • ________

= ________

GO ON

Find the volume of each figure.

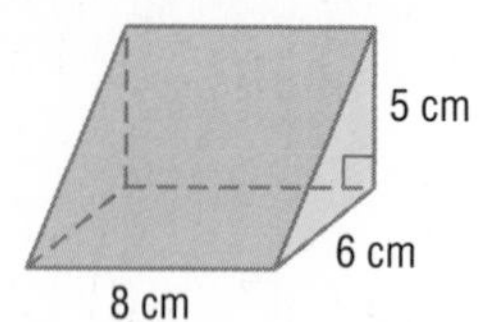

$B =$ ______

$B =$ ____(____ • ____) = ______

$V =$ ____ • ____ = ______

5

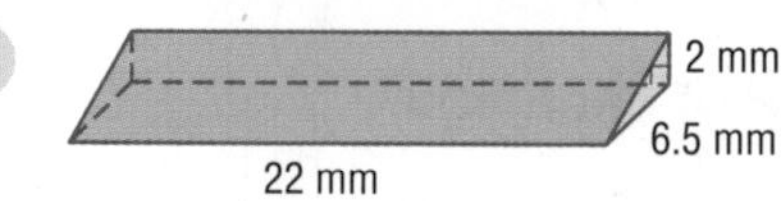

$B =$ ______

$B =$ ____(____ • ____) = ______

$V =$ ____ • ____ = ______

Step by Step Problem-Solving Practice

Solve.

6 **HOME IMPROVEMENT** Williams Home Solutions is installing a concrete sidewalk. The sidewalk's dimensions are 72 inches by 36 inches by 3 inches. How much cement will be needed?

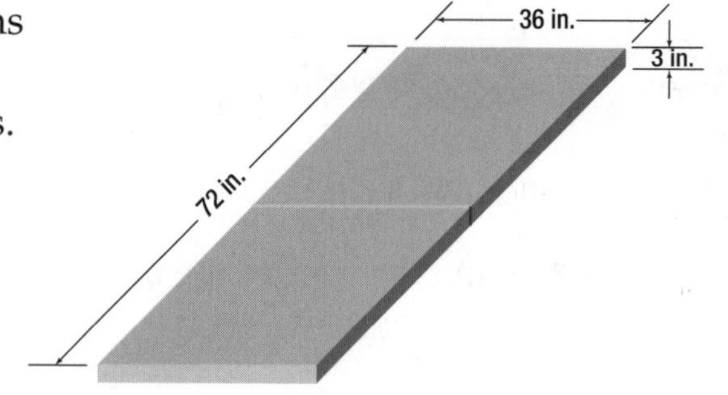

$B =$ ______ • ______

$B =$ ______ • ______ = ______

$V =$ ______ • ______

$V =$ ______ • ______ = ______

Check off each step.

______ **Understand: I underlined key words.**

______ **Plan: To solve the problem, I will** ____________________.

______ **Solve: The answer is** ____________________.

______ **Check: I checked my answer by** ____________________.

Skills, Concepts, and Problem Solving

Find the volume of each figure.

7 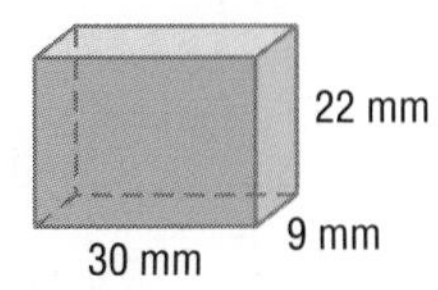

$V =$ ______________

$V =$ ______________

8 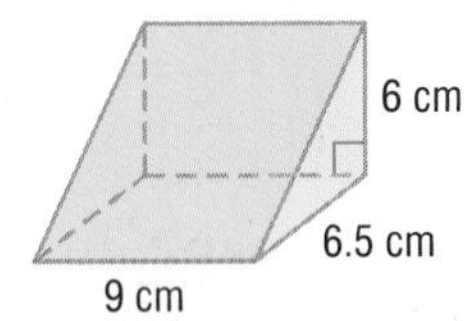

$V =$ ______________

$V =$ ______________

9 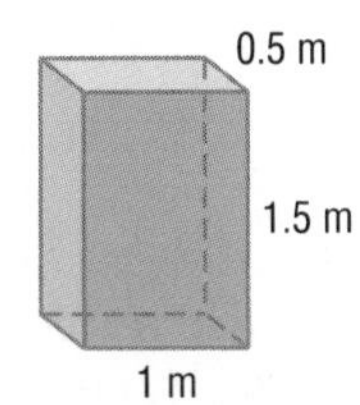

$V =$ ______________

$V =$ ______________

10 triangular prism

$b = 3$ cm, $h = 7$ cm,

$h_2 = 12$ cm

$V =$ ______________

$V =$ ______________

11 rectangular prism

$\ell = 5$ yd, $w = 2.5$ yd,

$h = 6$ yd

$V =$ ______________

$V =$ ______________

12 triangular prism

$b = 4$ in., $h = 8.5$ in.,

$h_2 = 7$ in.

$V =$ ______________

$V =$ ______________

Solve.

13 **MOVIES** Len bought a large box of popcorn at the movies. If the popcorn box measured 5 inches by 3 inches by 9 inches, what was the volume of Len's popcorn box?

____

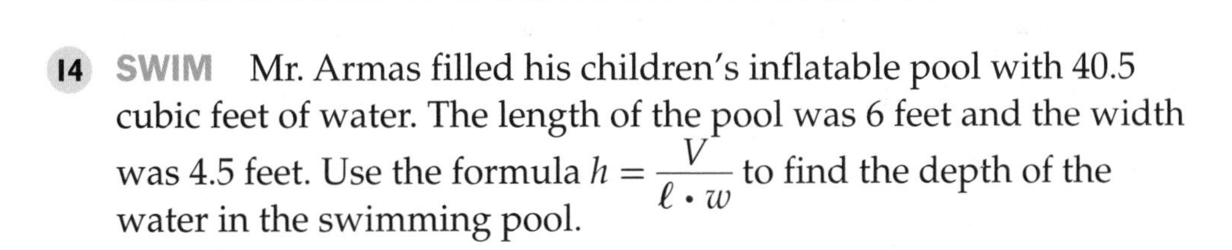

14 **SWIM** Mr. Armas filled his children's inflatable pool with 40.5 cubic feet of water. The length of the pool was 6 feet and the width was 4.5 feet. Use the formula $h = \frac{V}{\ell \cdot w}$ to find the depth of the water in the swimming pool.

____

GO ON

15 **TUNNELS** A triangular prism tunnel with dimensions 14 meters wide, 56 meters long, and 18 meters tall is to be built through a mountain. How many cubic meters of rock will need to be removed in order to complete the tunnel?

____

16 **CONSTRUCTION** To keep the wind from blowing his door shut, Brandon made a doorstop. The doorstop is shaped like a triangular prism. The base triangle has a length of 6 centimeters and a height of 15 centimeters. The height of the prism is 8 centimeters. What is the volume of Brandon's doorstop?

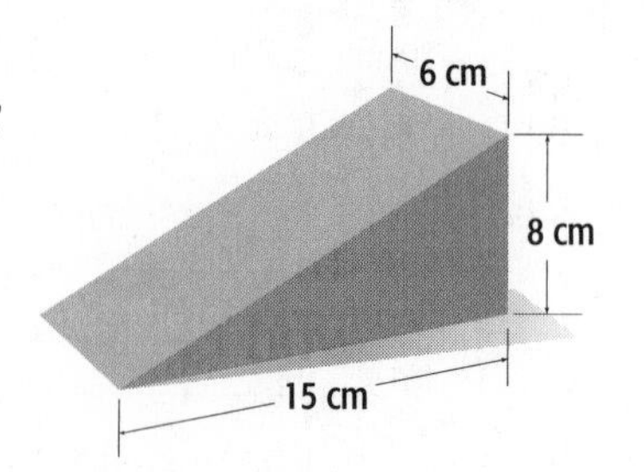

____

17 **MEASUREMENT** The Garrett family is building a pool in the shape of a rectangular prism in their backyard. The pool will cover an area 18 feet by 25 feet and will hold 2,700 cubic feet of water. If the pool is equal height throught, find the height.

____

18 **FOOD** Seri has a wedge of cheese that is in the shape of a triangular prism. The base of the triangle is 4 centimeters and the height of the triangle is 9 centimeters. The height of the whole wedge is 7 centimeters. What is the volume of Seri's wedge of cheese?

____

Vocabulary Check **Write the vocabulary word that completes each sentence.**

19 The faces on the top and bottom of a prism are called ____________.

20 A(n) ____________ is the unit used to measure volume.

21 A polyhedron with two parallel, congruent faces is called a ____________.

22 **Reflect** When you substitute values into the volume formula, can you interchange the base and height of a triangular prism and not affect its volume? Explain.

____

____

STOP

Chapter 5

Progress Check 3 (Lessons 5-5 and 5-6)

Find the area of each figure.

1

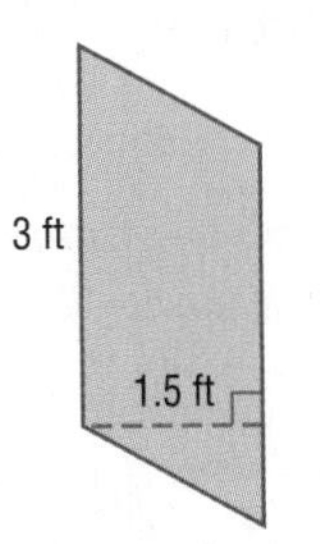

$A =$ ___________

2

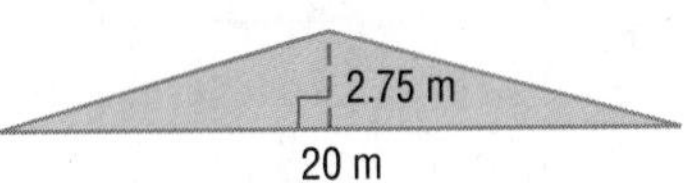

$A =$ ___________

3 triangle

$b = 10$ cm, $h = 12$ cm ___________

4 rectangle

$\ell = 12.5$ m, $w = 5$ m ___________

Find the volume of each figure.

5

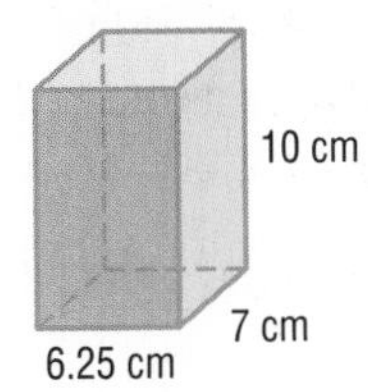

$V =$ ___________

6

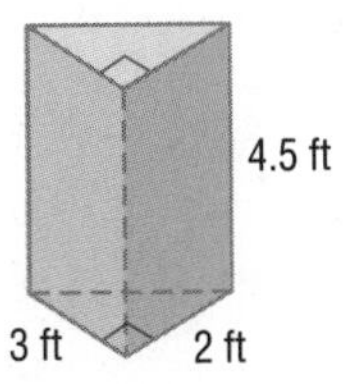

$V =$ ___________

7

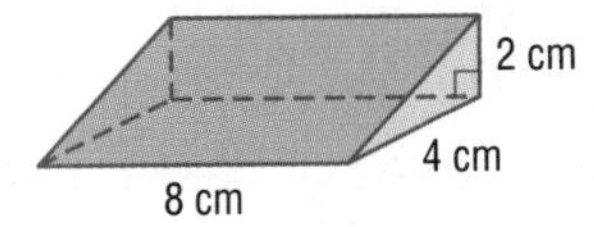

$V =$ ___________

8 triangular prism

$b = 4$ in., $h = 6$ in.,
$h_2 = 7.5$ in.

$V =$ ___________

9 rectangular prism

$\ell = 0.5$ yd, $w = 1.2$ yd,
$h = 3$ yd

$V =$ ___________

10 rectangular prism

$\ell = 16$ mm, $w = 25$ mm,
$h = 14$ mm

$V =$ ___________

Solve.

11 **PAPER GOODS** A full box of 50 tissues has a length of 8 inches and a height of 4.5 inches. If the width of the box is 3 inches, what is the volume of the box of tissue?

____

12 **LANDSCAPING** A landscaper is laying sod on a triangular piece of ground with a base length of 24 meters and a height of 45 meters. What is the area?

____

Lesson 5-7

Surface Area

KEY Concept

The **surface area** (SA) of a prism is the sum of the areas of each of its sides or faces. The **net** of a prism can help identify all the faces of a prism.

Rectangular Prism

$SA = 2\ell w + 2\ell h + 2hw$
$SA = 2(5 \cdot 3) + 2(5 \cdot 2) + 2(2 \cdot 3)$
$SA = 2(15) + 2(10) + 2(6)$
$SA = 30 + 20 + 12$
$SA = 62$ square units

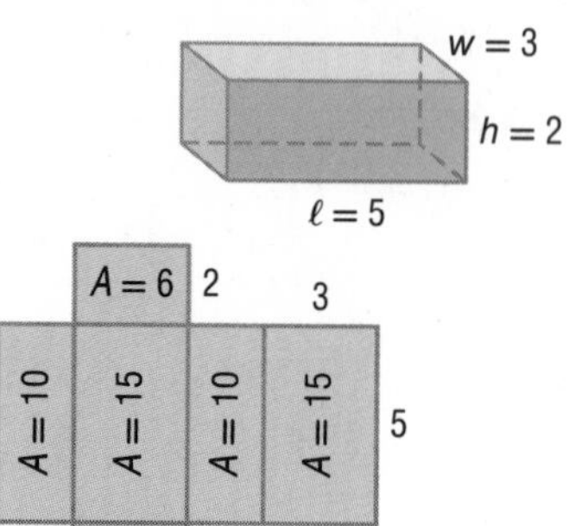

Triangular Prism

$SA =$ Area of bases + Area of the faces

$SA = 2(\frac{1}{2}bh) + \ell b + \ell w + \ell w$

$SA = 2(\frac{1}{2}(6 \cdot 4)) + (8 \cdot 6) + (8 \cdot 5) + (8 \cdot 5)$

$SA = 2(12) + 48 + 40 + 40$

$SA = 152$ square units

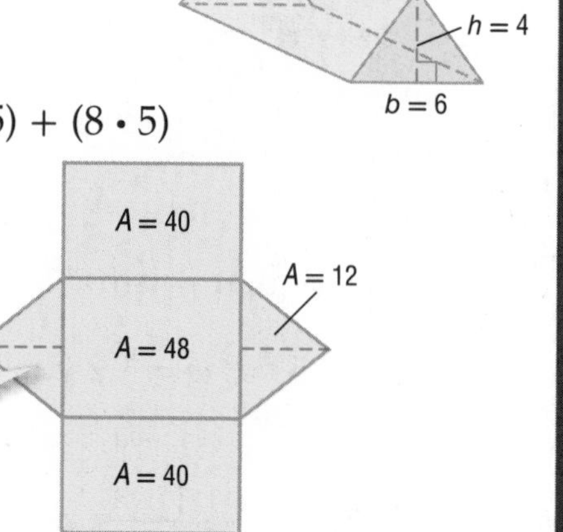

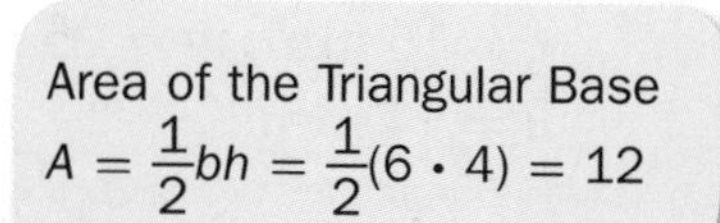

VOCABULARY

base
the faces on the top and bottom of a prism

face
the flat surface of a three-dimensional figure

net
a two-dimensional representation of a three-dimensional figure

surface area
the sum of the areas of all the faces of a three-dimensional figure

To find the surface area of a prism without a net, use area formulas.

Example 1

Draw a net to find the surface area of the prism.

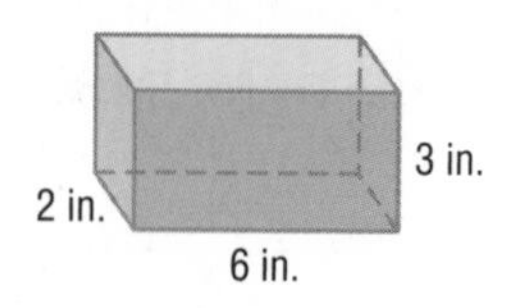

1. Draw a net. Label the dimensions of each region.
2. Use the formula for surface area.

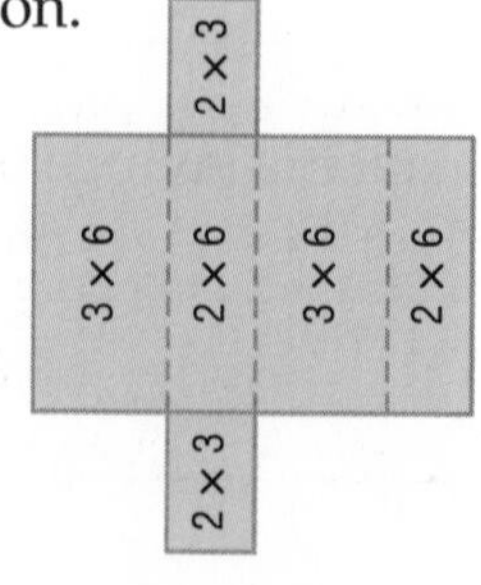

$SA = 2\ell w + 2\ell h + 2hw$
$SA = 2(2 \cdot 3) + 2(3 \cdot 6) + 2(2 \cdot 6)$
$SA = 2(6) + 2(18) + 2(12)$
$SA = 12 + 36 + 24$
$SA = 72 \text{ in}^2$

YOUR TURN!

Draw a net to find the surface area of the prism.

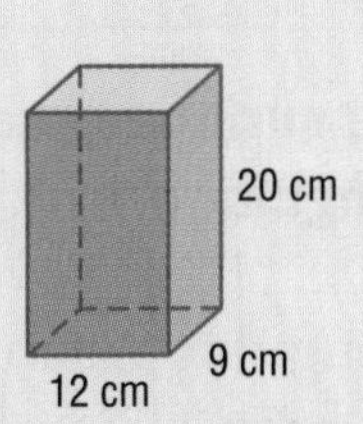

1. Draw a net. Label the dimensions of each region.
2. Use the formula for surface area.

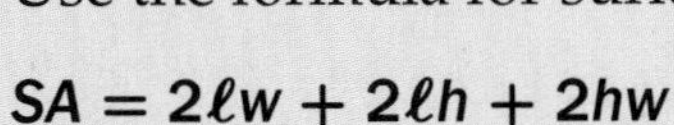

$SA = 2\ell w + 2\ell h + 2hw$

$SA = 2(\_\_\_\_ \cdot \_\_\_\_) + 2(\_\_\_\_ \cdot \_\_\_\_) + 2(\_\_\_\_ \cdot \_\_\_\_)$

$SA = 2(\_\_\_\_) + 2(\_\_\_\_) + 2(\_\_\_\_)$

$SA = \_\_\_\_ + \_\_\_\_ + \_\_\_\_$

$SA = \_\_\_\_\_\_$

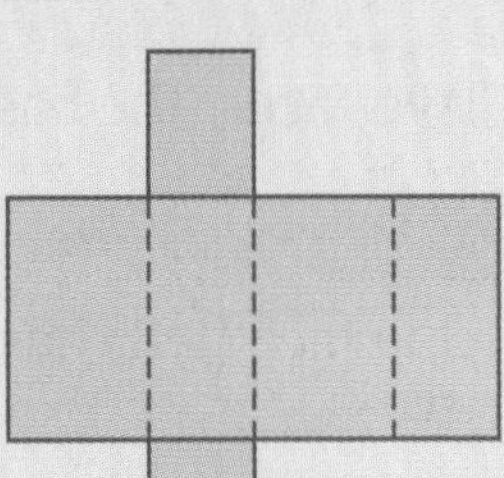

Example 2

Find the surface area of the rectangular prism.

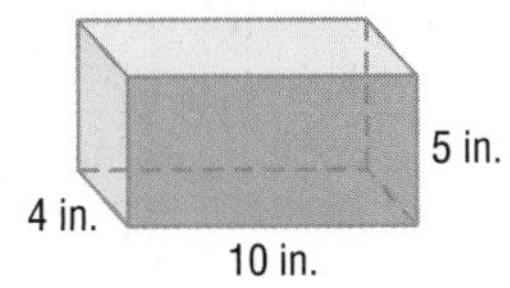

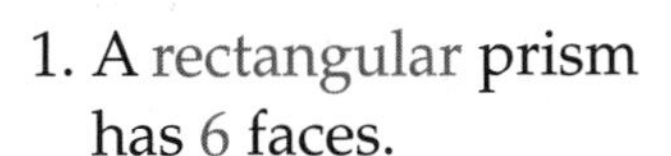

1. A rectangular prism has 6 faces.
2. Two faces (the bases) each have an area of $10 \cdot 4 = 40$.
3. Two faces (left and right) each have an area of $4 \cdot 5 = 20$.
4. Two faces (front and back) each have an area of $10 \cdot 5 = 50$.
5. Add the areas of all the faces.

   $40 + 40 + 20 + 20 + 50 + 50 = 220$
6. The rectangular prism has a surface area of 220 in².

YOUR TURN!

Find the surface area of the rectangular prism.

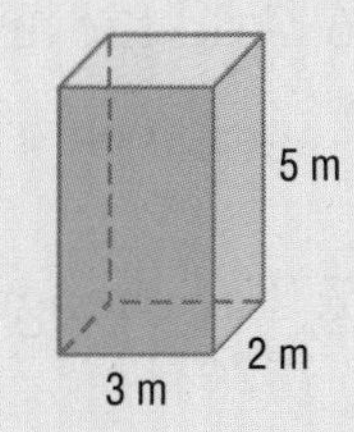

1. A ________ prism has ____ faces.
2. Two faces (the bases) each have an area of $\_\_\_\_ \cdot \_\_\_\_ = \_\_\_\_$.
3. Two faces (left and right) each have an area of $\_\_\_\_ \cdot \_\_\_\_ = \_\_\_\_$.
4. Two faces (front and back) each have an area of $\_\_\_\_ \cdot \_\_\_\_ = \_\_\_\_$.
5. Add the areas of all the faces.

   $\_\_\_ + \_\_\_ + \_\_\_ + \_\_\_ + \_\_\_ + \_\_\_ = \_\_\_$
6. The rectangular prism has a surface area of ________.

Example 3

Find the surface area of the triangular prism.

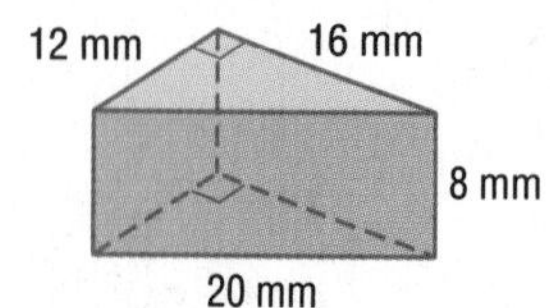

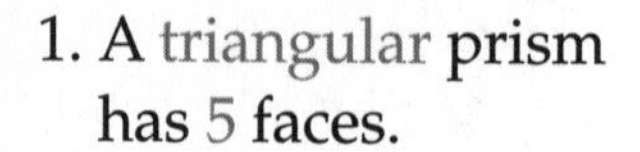

1. A triangular prism has 5 faces.
2. Two faces (the bases) each have an area of $\frac{1}{2}(12 \cdot 16) = 96$.
3. One face (left) has an area of $12 \cdot 8 = 96$.
4. One face (right) has an area of $16 \cdot 8 = 128$.
5. One face (bottom) has an area of $20 \cdot 8 = 160$.
6. Add the areas of all the faces.

   $96 + 96 + 96 + 128 + 160 = 576$

7. The triangular prism has a surface area of $576\ \text{mm}^2$.

YOUR TURN!

Find the surface area.

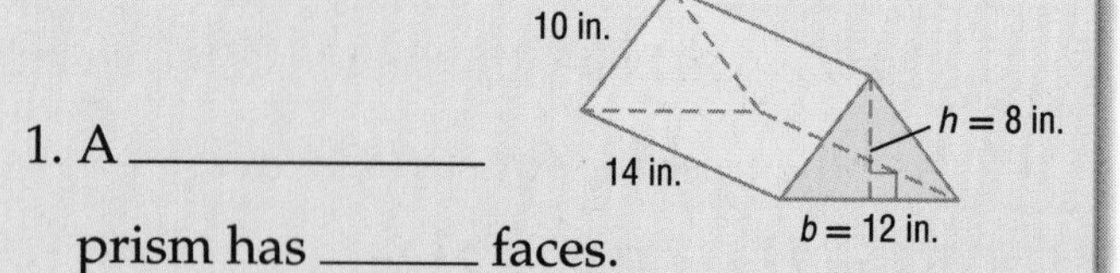

1. A ____________ prism has ______ faces.
2. Two faces (the bases) each have an area of $\frac{1}{2}(\_\_\_\_\_\_ \cdot \_\_\_\_\_\_) = \_\_\_\_\_\_$.
3. One face (left) has an area of

   ______ • ______ = ______.

4. One face (right) has an area of

   ______ • ______ = ______.

5. One face (bottom) has an area of

   ______ • ______ = ______.

6. Add the areas of all the faces.

   ____ + ____ + ____ + ____ + ____ = ____

7. The triangular prism has a surface area of

   ________.

Guided Practice

Draw a net to find the surface area of the prism.

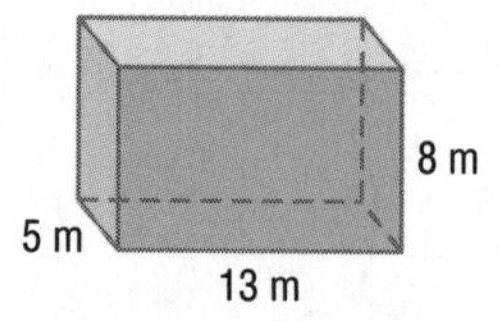

Use the formula for surface area.

$SA = 2\ell w + 2\ell h + 2hw$

$SA = 2(\_\_\_\_ \cdot \_\_\_\_) + 2(\_\_\_\_ \cdot \_\_\_\_) + 2(\_\_\_\_ \cdot \_\_\_\_)$

$SA = 2(\_\_\_\_) + 2(\_\_\_\_) + 2(\_\_\_\_)$

Find the sum of the areas.

________ + ________ + ________ = ________

Draw a net to find the surface area of the prism.

2

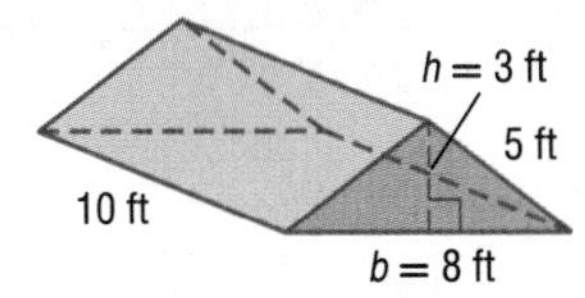

SA = Area of bases + Area of the faces

$SA = 2(\frac{1}{2}bh) + \ell b + \ell w + \ell w$

SA = 2 ____(____ • ____) + (____ • ____) + (____ • ____) + (____ • ____)

SA = 2(____) + ____ + ____ + ____

SA = ________

Find the surface area of each rectangular prism.

3

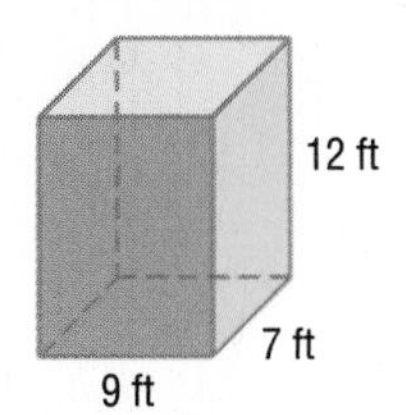

area of two bases

A = ______ = 2(____ • ____) = ______ ft²

area of two faces (top and bottom)

A = ______ = 2(____ • ____) = ______ ft²

area of two faces (right and left)

A = ______ = 2(____ • ____) = ______ ft²

SA = _____ + _____ + _____ = _______

4

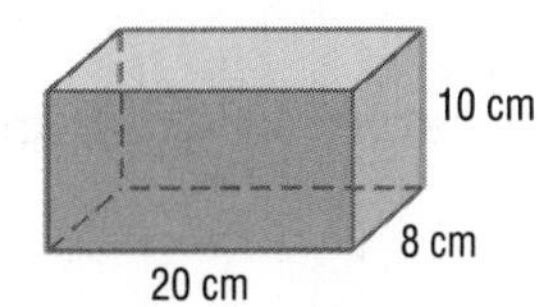

area of two bases

A = ______ = 2(____ • ____) = ______ ft²

area of two faces (top and bottom)

A = ______ = 2(____ • ____) = ______ ft²

area of two faces (right and left)

A = ______ = 2(____ • ____) = ______ ft²

SA = _____ + _____ + _____ = _______

GO ON

Step by Step Practice

5 Find the surface area of the triangular prism.

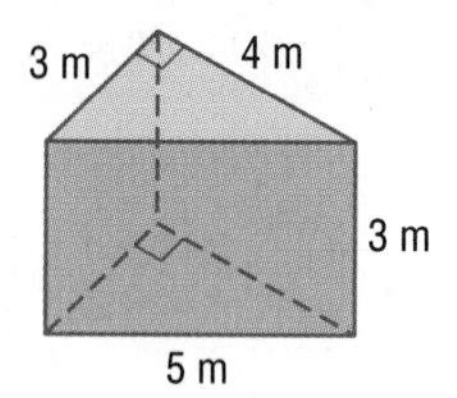

Step 1 Find the area of the two bases.

$A = 2\left(\_\_\_\_\right) = \_\_\_\_ (\_\_\_\_ \cdot \_\_\_\_)$

$= 2 \cdot \_\_\_\_ = \_\_\_\_ \text{ m}^2$

Step 2 Find the area of the left face. $A = \_\_\_\_ \cdot \_\_\_\_ = \_\_\_\_ \text{ m}^2$

Step 3 Find the area of the right face. $A = \_\_\_\_ \cdot \_\_\_\_ = \_\_\_\_ \text{ m}^2$

Step 4 Find the area of the bottom face. $A = \_\_\_\_ \cdot \_\_\_\_ = \_\_\_\_ \text{ m}^2$

Step 5 Add the areas of all the faces. $SA = \_\_\_\_ + \_\_\_\_ + \_\_\_\_ + \_\_\_\_ = \_\_\_\_$

Find the surface area of each triangular prism.

6

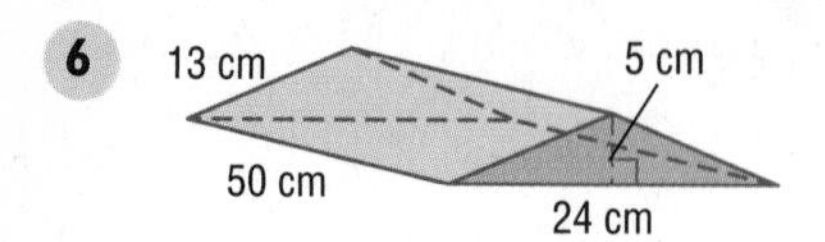

area of two bases

$SA = 2\left(\_\_\_\_\right) = \_\_\_\_ (\_\_\_\_ \cdot \_\_\_\_)$

$= 2 \cdot \_\_\_\_ = \_\_\_\_ \text{ cm}^2$

area of the left face

$SA = \_\_\_\_ \cdot \_\_\_\_ = \_\_\_\_ \text{ cm}^2$

area of the right face

$SA = \_\_\_\_ \cdot \_\_\_\_ = \_\_\_\_ \text{ cm}^2$

area of the bottom face

$SA = \_\_\_\_ \cdot \_\_\_\_ = \_\_\_\_ \text{ cm}^2$

Add the areas of all the faces.

$\_\_\_\_ + \_\_\_\_ + \_\_\_\_ + \_\_\_\_ = \_\_\_\_$

7

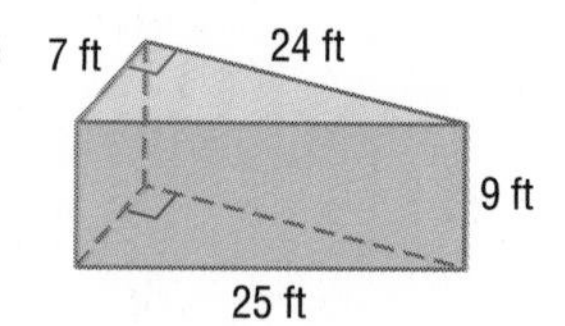

area of two bases

$SA = 2\left(\_\_\_\_\right) = \_\_\_\_ (\_\_\_\_ \cdot \_\_\_\_)$

$= 2 \cdot \_\_\_\_ = \_\_\_\_ \text{ ft}^2$

area of the left face

$SA = \_\_\_\_ \cdot \_\_\_\_ = \_\_\_\_ \text{ ft}^2$

area of the right face

$SA = \_\_\_\_ \cdot \_\_\_\_ = \_\_\_\_ \text{ ft}^2$

area of the bottom face

$SA = \_\_\_\_ \cdot \_\_\_\_ = \_\_\_\_ \text{ ft}^2$

Add the areas of all the faces.

$\_\_\_\_ + \_\_\_\_ + \_\_\_\_ + \_\_\_\_ = \_\_\_\_$

Step by Step Problem-Solving Practice

Solve.

8 **PAINTING** Spencer is painting his bedroom door. If the dimensions of his door are 84 inches tall by 40 inches wide by 2.5 inches thick, how much surface area will the paint need to cover?

$SA = 2\ell w + 2\ell h + 2hw$

$SA = 2(\_\_\_\_ \cdot \_\_\_\_) + 2(\_\_\_\_ \cdot \_\_\_\_) + 2(\_\_\_\_ \cdot \_\_\_\_)$

$SA = 2(\_\_\_\_) + 2(\_\_\_\_) + 2(\_\_\_\_)$

$SA = \_\_\_\_ + \_\_\_\_ + \_\_\_\_$

$SA = \_\_\_\_\_\_$

Check off each step.

____ **Understand: I underlined key words.**

____ **Plan: To solve the problem, I will** ____________.

____ **Solve: The answer is** ____________.

____ **Check: I checked my answer by** ____________.

Skills, Concepts, and Problem Solving

Find the surface area of each rectangular prism.

9 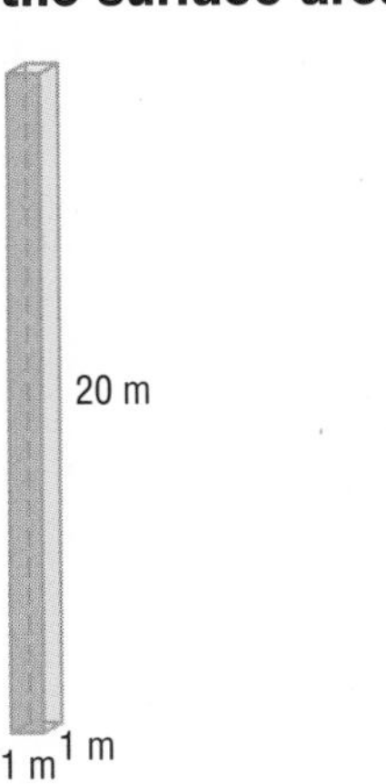

$SA = \_\_\_\_\_\_$

10 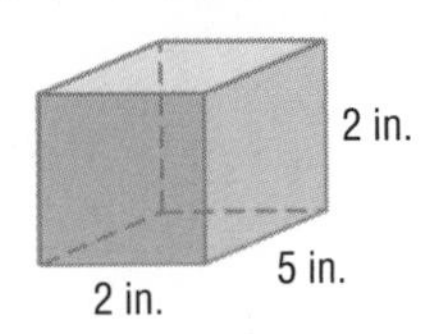

$SA = \_\_\_\_\_\_$

11 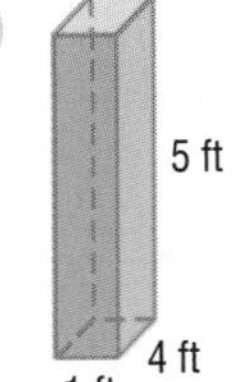

$SA = \_\_\_\_\_\_$

Find the surface area of each prism.

12 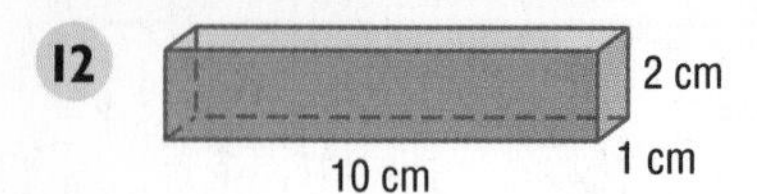

$SA =$ ______________

13

h = 6 yd
10 yd
20 yd
16 yd

$SA =$ ______________

14 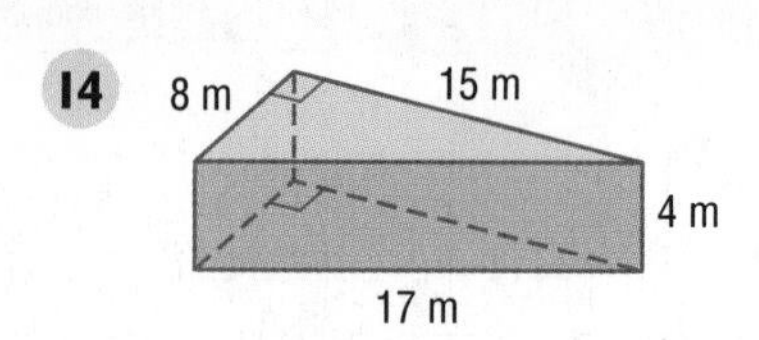

$SA =$ ______________

Solve.

CONSTRUCTION Yolanda has a sheet of wood with dimensions of 10 feet by 15 feet. She uses a saw and duct tape to make the entire sheet of wood into a cube. (A cube is a rectangular prism with six faces that are congruent squares.)

15 What is the surface area of the box? ______________

16 What is the surface area of one face? ______________

17 What is the length of each edge of the box? ______________

Vocabulary Check **Write the vocabulary word that completes each sentence.**

18 A(n) ______________ is a two-dimensional representation of a three-dimensional figure.

19 The flat surface of a three-dimensional figure is called a(n)

______________.

20 The faces on the top and bottom of a prism are the

______________.

21 A solid figure with two congruent parallel bases is a(n)

______________.

22 **Reflect** Name two differences between finding the surface area of a rectangular prism and a triangular prism.

____

____

____

____

STOP

Lesson 5-8

The Coordinate Plane

KEY Concept

A **coordinate plane** is a grid made of two intersecting number lines. These lines are called the x-axis and the y-axis.

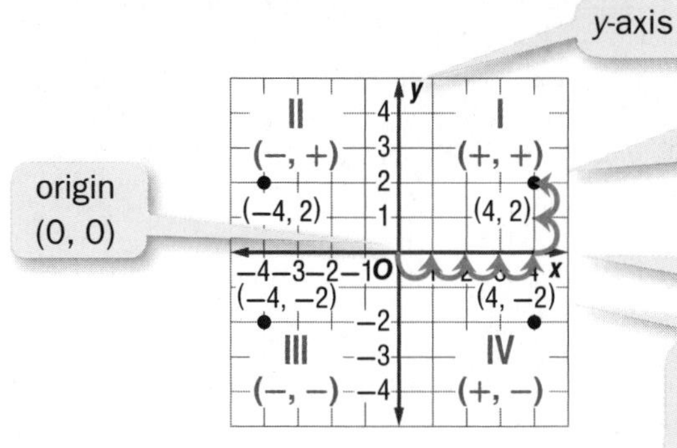

The **x-axis** is the vertical number line and the **y-axis** is the horizontal number line. An **ordered pair** (x, y) gives the location of a point on the coordinate plane.

The coordinate plane is divided into 4 quadrants: I, II, III, and IV. A point lies in one of the quadrants or on an axis.

VOCABULARY

coordinate plane
a plane in which a horizontal number line and a vertical number line intersect at their zero points

ordered pair
a pair of numbers used to locate a point in the coordinate plane

origin
the point (0, 0) on a coordinate graph

x-axis
the horizontal axis on a coordinate plane

y-axis
the vertical axis on a coordinate plane

In an ordered pair, x comes before y, just as x comes before y in the alphabet.

Example 1

Write the ordered pair for the point. Name the quadrant in which the point is located.

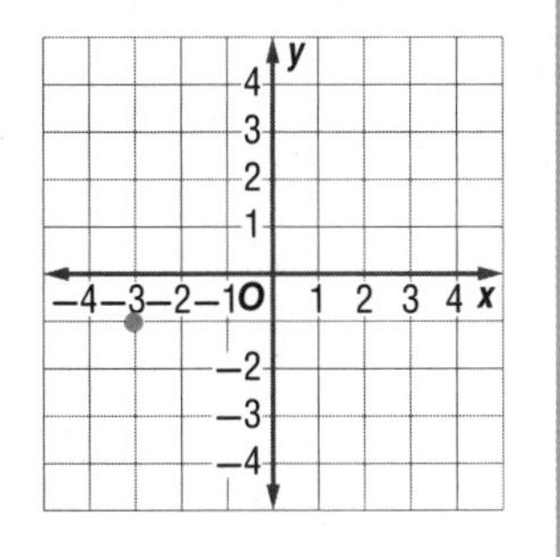

1. Begin at the origin.
2. Move 3 unit(s) left along the x-axis.
3. Move 1 unit(s) down along the y-axis.
4. The ordered pair is (−3, −1).
5. The x-coordinate is negative and the y-coordinate is negative. The point is in Quadrant III.

YOUR TURN!

Write the ordered pair for the point. Name the quadrant in which the point is located.

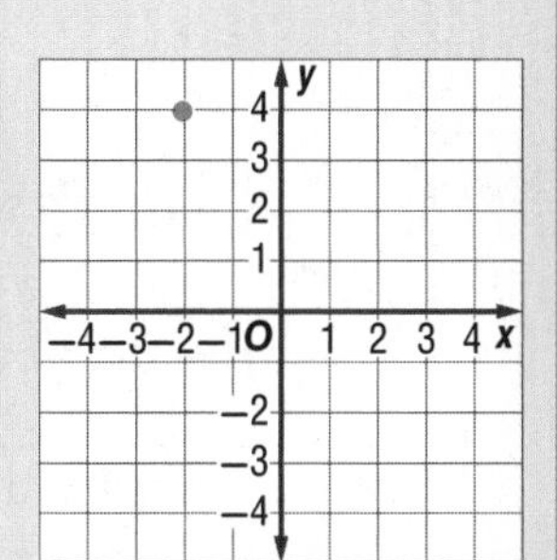

1. Begin at the origin.
2. Move ____ unit(s) ________ along the x-axis.
3. Move ____ unit(s) ____ along the y-axis.
4. The ordered pair is (__________).
5. The x-coordinate is __________ and the y-coordinate is __________. The point is in Quadrant __________.

GO ON

Example 2

Graph the point (1, 0). Name the axis or quadrant where the point is located.

1. Begin at the origin. The x-coordinate is 1. Move 1 unit to the right.

2. The y-coordinate is 0. Move 0 units up or down.

3. Draw the point and label it.

4. The point is on the x-axis.

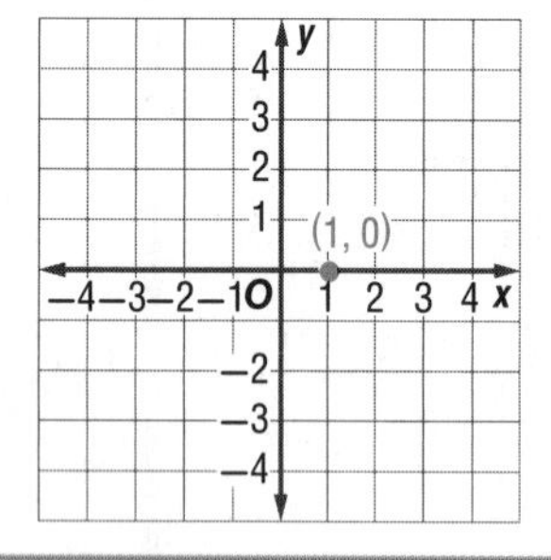

YOUR TURN!

Graph the point (0, −2). Name the axis or quadrant where the point is located.

1. Begin at the origin. The x-coordinate is _____. Move _____ units ____________.

2. The y-coordinate is _____. Move _____ unit(s) ____________.

3. Draw the point and label it.

4. The point is ____________.

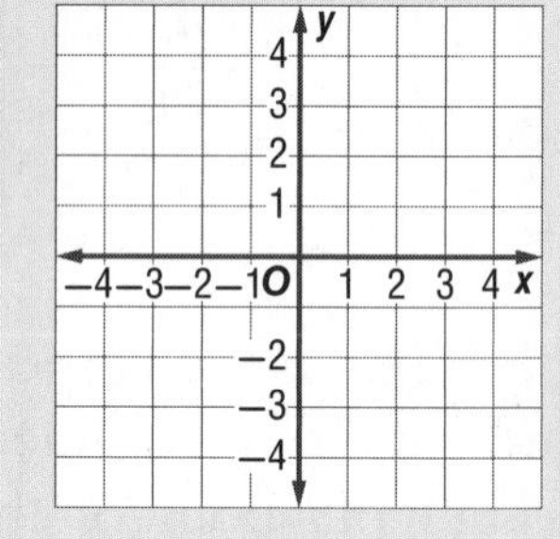

Guided Practice

Write the ordered pair for each point. Name the axis or quadrant in which each point lies.

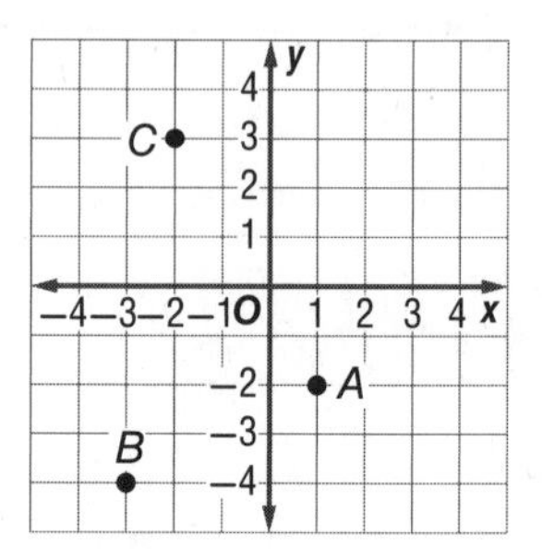

1. A ____________ ____________
2. B ____________ ____________
3. C ____________ ____________

Step by Step Practice

4. Graph the point (3, −2).

Step 1 Begin at the origin.
The x-coordinate is _____.

Move _____ unit(s) to the _____.

Step 2 The y-coordinate is _____.

Move _____ units _____.

Step 3 Draw the point and label it.

Step 4 The point is ____________.

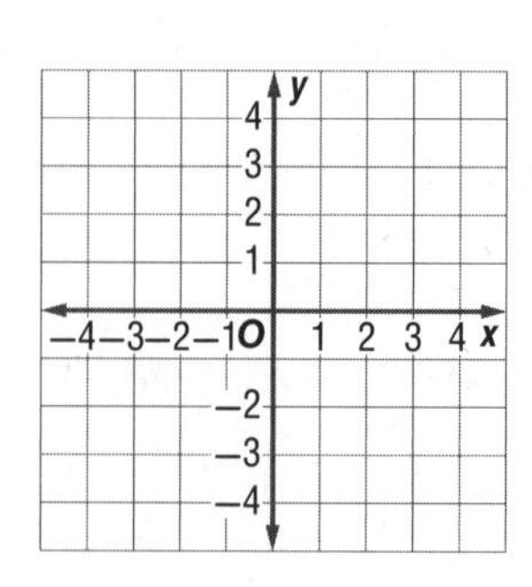

Graph each point on the graph to the right.

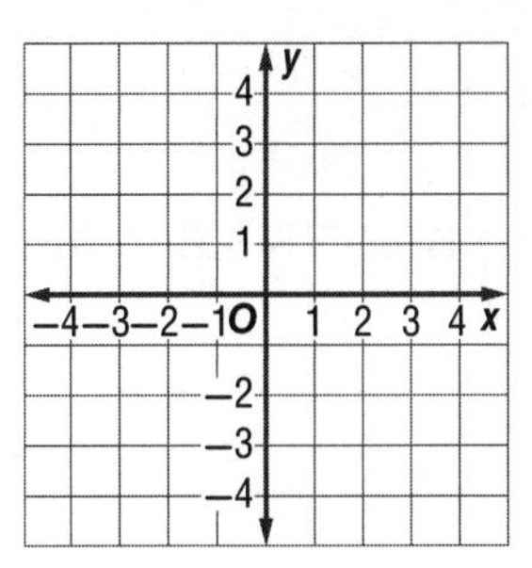

5 $D(-1, -1)$

6 $F(0, -3)$

7 A point in the fourth quadrant. Name the point G.

8 A point in the first quadrant. Name the point H.

Step by Step Problem-Solving Practice

Solve.

9 **WALKING** Jerome walks from his house to Patrick's house. Jerome's house is located at (5, –3). If he walks 7 blocks north and 5 blocks west, at what point is Patrick's house located?

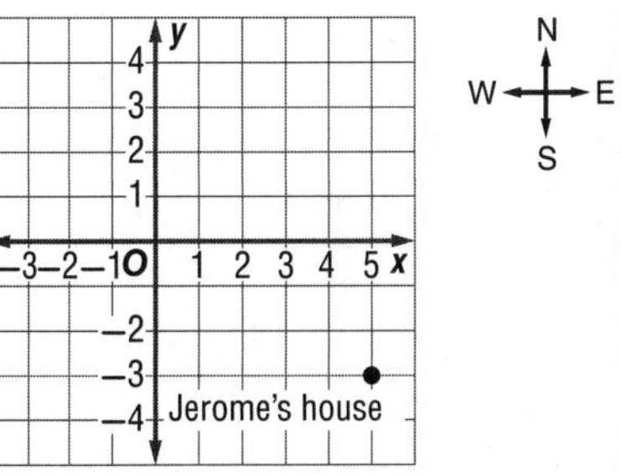

Check off each step.

_____ **Understand: I underlined key words.**

_____ **Plan: To solve the problem, I will** ________________________.

_____ **Solve: The answer is** ________________________.

_____ **Check: I checked my answer by** ________________________

________________________.

Skills, Concepts, and Problem Solving

Graph each point. Name the axis or quadrant where the point is located.

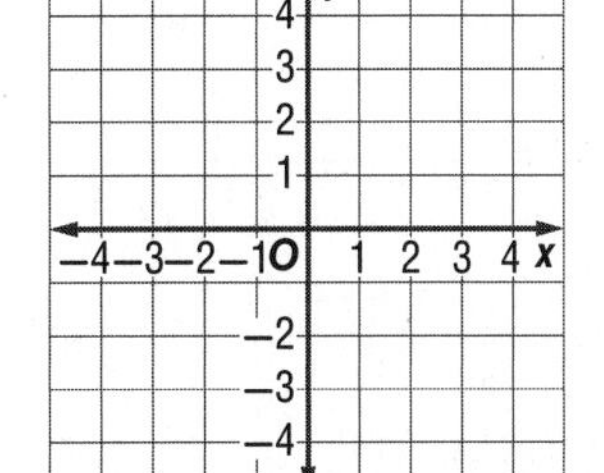

10 $J(0, 3)$ ____________

11 $K(-3, 4)$ ____________

12 $L(1, -1)$ ____________

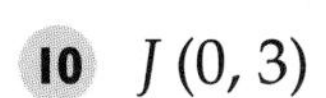

13 $M(4, 1)$ ____________

14 $N(-2, 0)$ ____________

15 $P(-3, -2)$ ____________

Solve.

16 **GEOMETRY** Victor placed four ordered pairs, (–4, 5), (4, 5), (–4, –2), and (4, –2), on a coordinate plane. When he connected the points, what shape did he make?

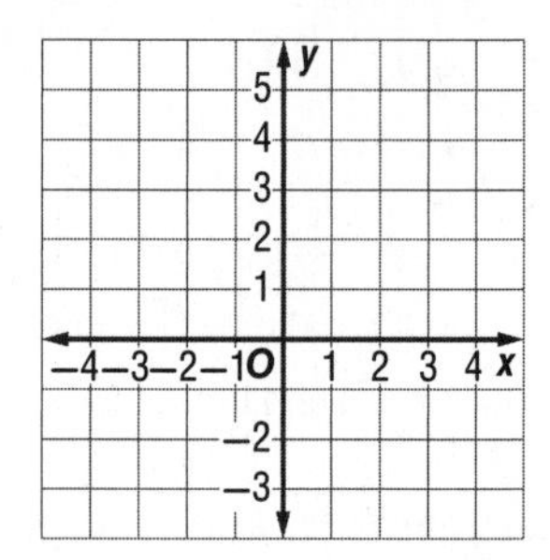

____

17 **GEOMETRY** Terri placed four ordered pairs, (–2, 1), (3, 2), (–3, –1), (2, 0), on a coordinate plane. When she connected the points, what shape did she make?

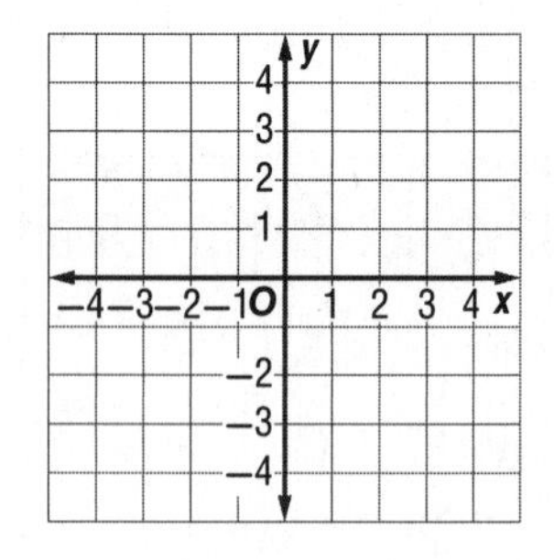

____

18 **GEOMETRY** Juan drew a rectangle on a coordinate plane. The length of the rectangle is 7 units and the width is 3 units. One corner or vertex is located at the origin. What could be the coordinates of the other 3 vertices?

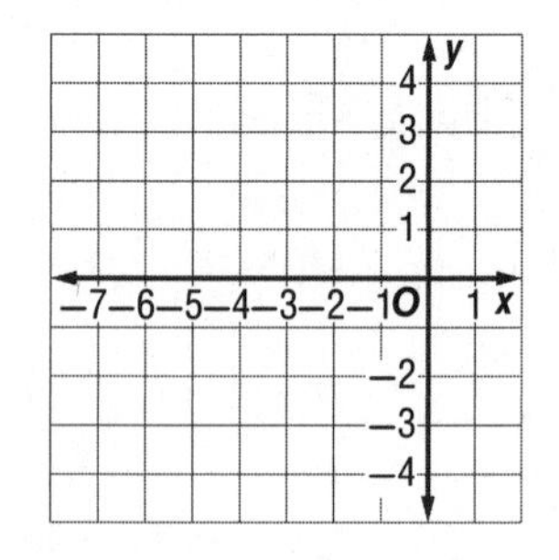

____

Vocabulary Check **Write the vocabulary word that completes each sentence.**

19 A(n) ____________________ is a grid in which a horizontal number line and a vertical number line intersect at their zero points.

20 The ____________________ is the point (0, 0) on a coordinate graph.

21 The vertical axis on a coordinate plane is the ____________________.

22 **Reflect** Consider the ordered pair (4, –6). Without using a graph, explain how to determine its location on the coordinate plane.

____

____

____

STOP

Chapter 5

Progress Check 4 (Lessons 5-7 and 5-8)

Find the surface area of each prism.

1

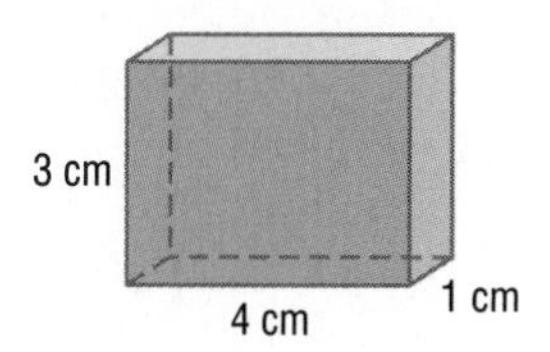

$SA =$ ____________________

2

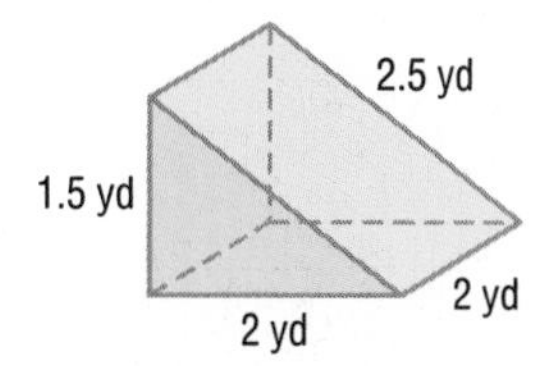

$SA =$ ____________________

Graph each point. Name the axis or quadrant where the point is located.

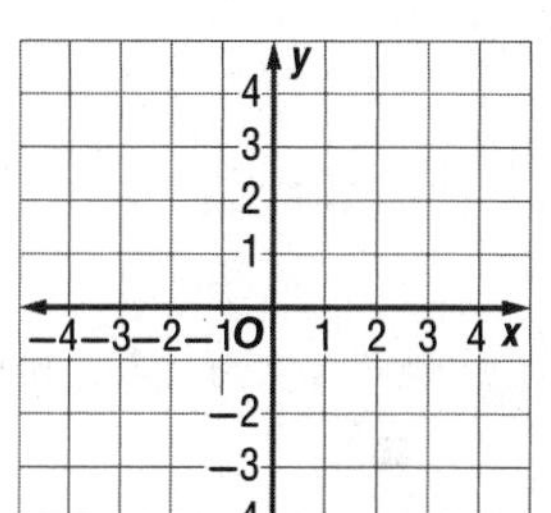

3 $R(-4, -4)$

____

4 $S(0, -4)$

____

5 $T(0, 2)$

____

6 $V(1, 1)$

____

7 $W(5, -3)$

____

8 $Z(4, 0)$

____

Solve.

PET CARE Kendra has wire fencing that has the dimensions 9 feet by 6 feet. She uses the wire to make a cube-shaped crate for her puppy.

9 What is the surface area of the cage?

____

10 What is the area of one face?

____

11 What is the length of each edge of the cage?

____

12 **GEOMETRY** Denise drew a rectangle on a coordinate plane that covers the origin. The length of the rectangle is 12 units and the width is 5 units. One corner or vertex is located at the (−5, 0) and (−5, −5). What are the coordinates of the other two vertices?

____

Chapter 5

Chapter 5 Test

Complete each conversion.

1. 100 pt = ______________ gal
2. 12 c = ______________ oz
3. 50 m = ______________ cm
4. 180 mm = ______________ m
5. 40 L = ______________ kL
6. 67 mg = ______________ kg

Complete each conversion. Round to the nearest thousandth.

7. 150 mL ≈ ______________ fl oz
8. 15 L ≈ ______________ qt
9. 160 g ≈ ______________ oz
10. 72 m ≈ ______________ yd

Find the perimeter of each figure.

11. 12 m, 12 m

$P =$ ______________

12. 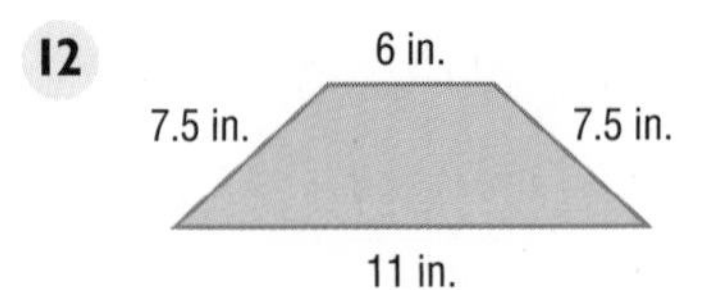

$P =$ ______________

Find the area of each figure.

13. 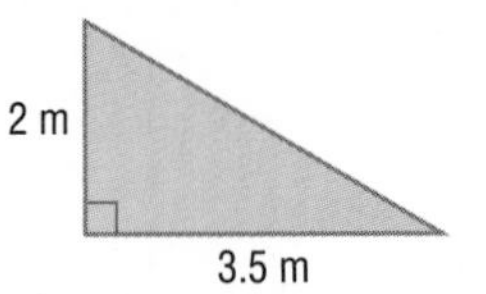

$A =$ ______________

14. 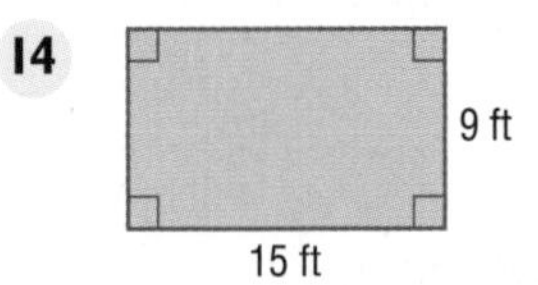

$A =$ ______________

Find the volume and surface area of each prism.

15. 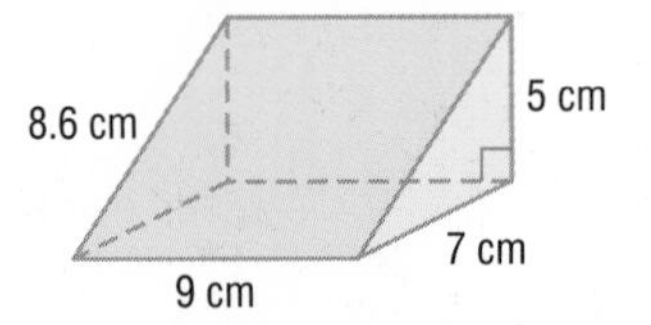

$V =$ ______________

$SA =$ ______________

16. 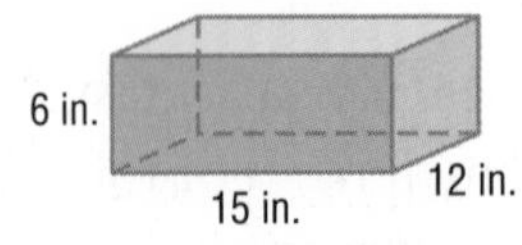

$V =$ ______________

$SA =$ ______________

Graph each point. Name the axis or quadrant where the point is located.

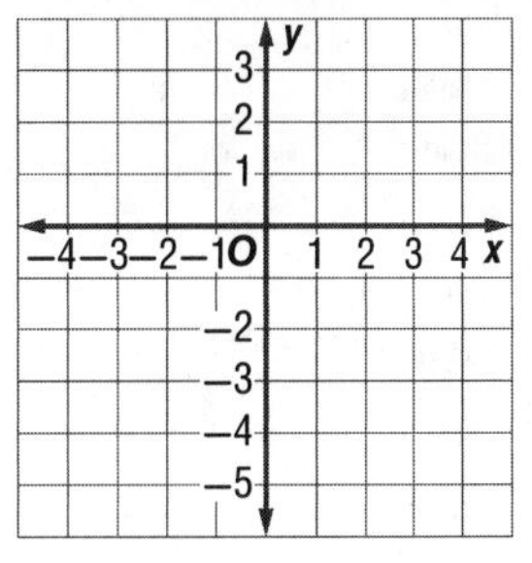

17 $A\ (0, -5)$ ________

18 $B\ (1, -2)$ ________

19 $C\ (3, 2)$ ________

20 $D\ (4, -1)$ ________

21 $F\ (2, 0)$ ________

22 $G\ (-1, -1)$ ________

Solve.

23 **SAILING** A triangular sail has a height of 3 meters and a base of 2.5 meters. What is the area of the sail?

____

24 **ART** Natalie uses acrylic paints on a rectangular canvas with an area of 13.5 square feet. If the canvas has a length of 3 feet, what is the height?

____

25 **VOLUME** A perfume bottle is in the shape of a triangular prism. The volume of the bottle is 30 centimeters cubed. If the area of the base is 6 square centimeters, what is the height of the bottle?

____

26 **PACKAGING** Bottle A holds 1.75 L of oil. Bottle B holds 0.002 kL of oil. Which bottle holds more oil?

____

Correct the mistake.

27 **HEALTH** Oscar lost 12 kilograms in the last year. He told his friend that he lost 5.45 lb. What miscalculation did Oscar do when he converted the kilograms to pounds? About how many pounds did Oscar lose?

____

____

STOP

Probability and Statistics

Speed

Mean, median, and mode are used to describe sets of data. For example, the mean, or average speed of the winning car in the 2008 Indianapolis 500 was 143.567 mph.

STEP 1 Chapter Pretest Are you ready for Chapter 6? Take the Chapter 6 Pretest to find out.

STEP 2 Preview Get ready for Chapter 6. Review these skills and compare them with what you will learn in this chapter.

What You Know

You know how to multiply by powers of ten.

Example:

$14.5 \cdot 100$

$= 1450$

To multiply 14.5 · 100, move the decimal point 2 places to the right.

TRY IT!

Multiply.

1. $16.8 \cdot 10 =$ ________
2. $33.42 \cdot 100 =$ ________

What You Will Learn

Lesson 6-2

To change a metric measure from one unit to another, you can use the relationship between the two units and multiply by a power of 10.

Example:

Convert 3.6 meters to centimeters.

Use the relationship 1 m = 100 cm.

$3.6 \cdot 100 = 360$

$3.6 \text{ m} = 360 \text{ cm}$

What You Know

You know how to order a set of numbers from least to greatest.

Example: 4, 15, 2, 8, 14

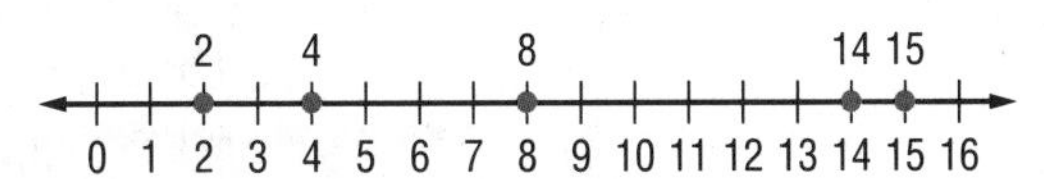

The numbers from least to greatest are 2, 4, 8, 14, 15.

What You Will Learn

Lesson 6-5

The **median** of a set of data is the middle number of data that has been written in order.

Example:

ages of students: 14, 12, 15, 13, 13

in order: 12, 13, 13, 14, 15

The median is 13.

What You Know

You can find the value of the following expression.

$(8 + 9 + 13) \div 3 = (30) \div 3$

$= 10$

TRY IT!

Simplify.

3. $(16 + 8) \div 2 =$ ________
4. $(10 + 5 + 22 + 17) \div 4 =$ ________

What You Will Learn

Lesson 6-5

The **mean** of a set of data is the sum of the data divided by the number of pieces of data.

Example:

data set: 1, 3, 5, 7

$$\text{mean} = \frac{1 + 3 + 5 + 7}{4} = \frac{16}{4} = 4$$

Line Graphs

A **line graph** shows how **data** changes over time.

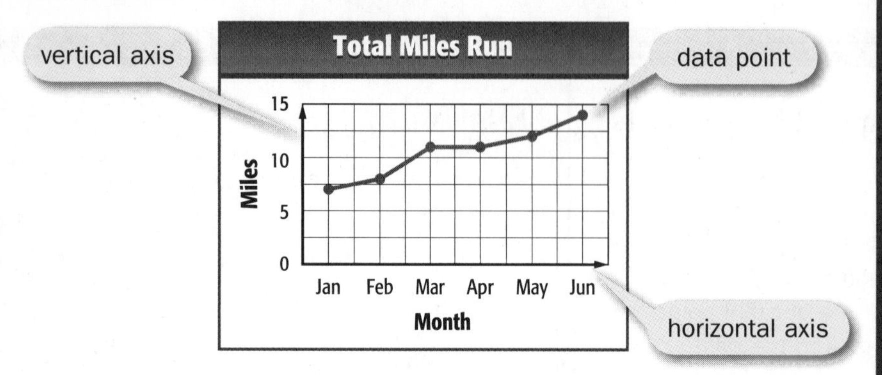

This graph shows that as the months increase the number of miles run increase.

VOCABULARY

data
pieces of information, which are often numerical

horizontal axis
the axis on which the categories or values are shown in a bar and line graph; the *x*-axis

line graph
a graph that shows how a set of data changes over a period of time

vertical axis
the axis on which the scale and interval are shown in a bar and line graph; the *y*-axis

A line graph has a title and labeled axes. You can analyze data in a line graph by reading the title and labels along with observing the relationship of the data points.

Example 1

Use the line graph to make a table of the data.

1. The horizontal axis represents the year. Write this in the table.
2. The vertical axis represents the value of comic books. Write this in the table.
3. Start on the horizontal axis and locate Year 1. Move up to the point above Year 1. Read the vertical scale to find the value of the comic books.
4. This point gives you the data Year 1, $2000.
5. Repeat Steps 3 and 4 to gather the data for Years 2 through 6.

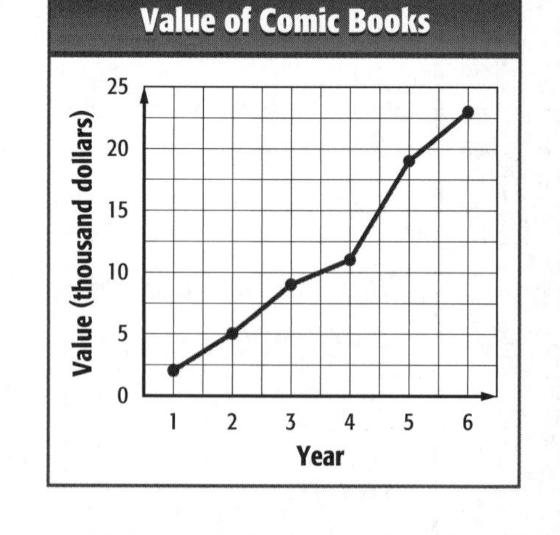

Comic Book Values						
Year	1	2	3	4	5	6
Value (thousand dollars)	2	5	9	11	19	23

YOUR TURN!

Use the line graph to make a table of the data.

1. The horizontal axis represents the __________.
   Write this in the table.

2. The vertical axis represents the number of __________.
   Write this in the table.

3. On the horizontal axis, locate __________

   Move up to the point above __________

   Read the vertical scale to find the number

   of __________.

4. This point gives you the data

   __________, __________.

5. Repeat Steps 3 and 4 to gather the data for the other times.

Calls per Hour

Number of Calls: 30, 25, 20, 15, 10, 5, 0

8:00 A.M. 9:00 A.M. 10:00 A.M. 11:00 A.M. 12:00 P.M. 1:00 P.M.

Time

____

Example 2

Make a line graph of the data. Describe the change in data.

1. Draw and label the axes. The horizontal axis is the year. Write each year along the axis.

2. The vertical axis is the depth. Use a scale of 5.

3. Title the graph *Average Water Depth.*

4. Plot each data point. The first point is (2003, 25). Connect each point with a straight line.

5. The average depth increases, and then decreases as the years increase.

Average Depth of Water					
Year	2003	2004	2005	2006	2007
Depth (in meters)	25	26	22	21	20

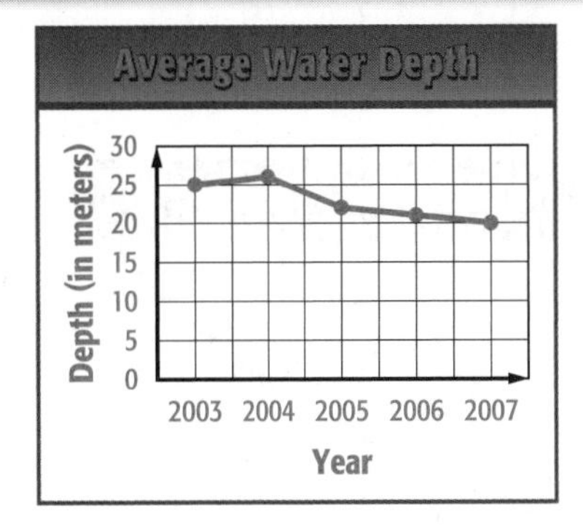

GO ON

YOUR TURN!

Make a line graph of the data. Describe the change in data.

Plant Growth					
Month	May	June	July	Aug	Sept
Height (in inches)	45	49	65	76	77

1. Draw and label the axes. The horizontal axis is the ________. Write each month along the axis.
2. The vertical axis is the ________. Use a scale of ________.
3. Title the graph ______________.
4. Plot each data point. The first point is (May, ________). Connect each point with a ______________.
5. The height ______________ as the months increase.

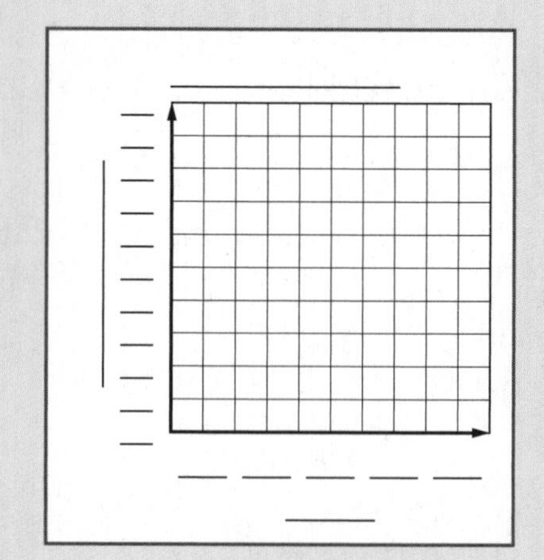

Guided Practice

Use the line graph to make a table of the data.

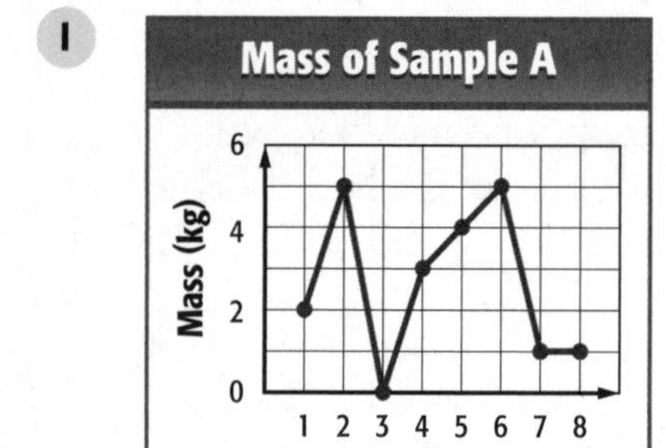

____

Week	1							
Mass (kg)	2							

2

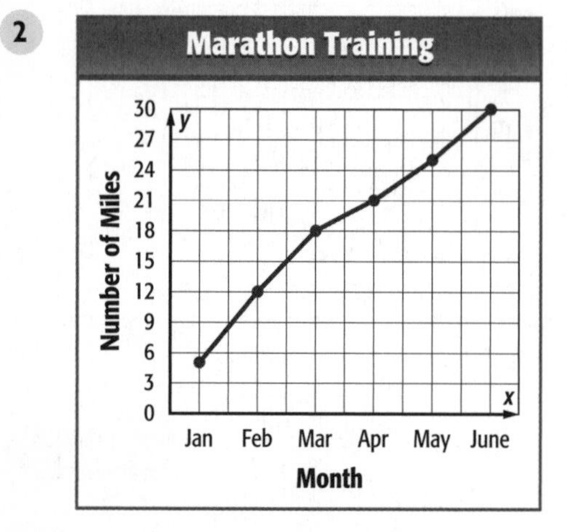

____

Month	Miles

Step by Step Practice

3 The table below shows the height of a plant each day. Make a line graph of the data. Describe the change in data.

Day	1	2	3	4	5	6	7
Height (cm)	0.5	1	1.3	2.2	2.6	3	3.5

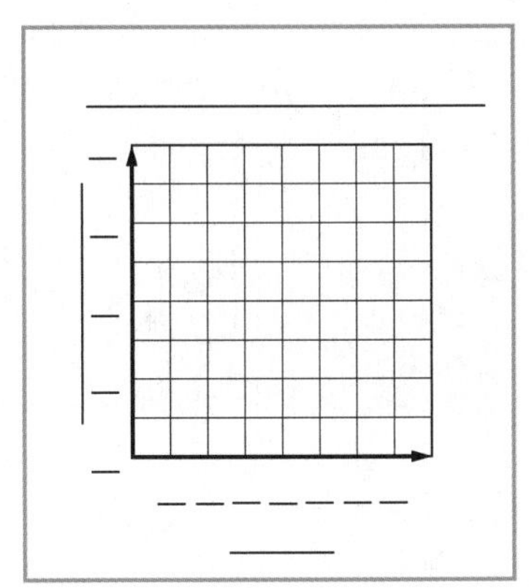

Step 1 Draw and label the axes.

The horizontal axis is the ______.
Write the days 1 to 7 along the axis.

Step 2 The vertical axis is the ____________________.
Use a scale of 1.

Step 3 Title the graph.

Step 4 Plot each data point. The first point is (Day 1, ______).
Connect each point to the next point with a

____________________.

Step 5 The plant height ____________________ as the days increase.

4 The table below shows the price of first-class stamps in certain years. Make a line graph of the data. Describe the change in data.

Year	1970	1980	1990	2000
Price ($)	$0.06	$0.15	$0.25	$0.33

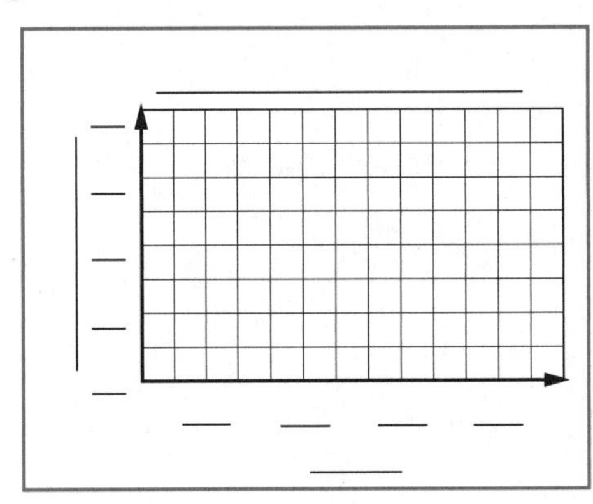

____

____

____

GO ON

Step by Step Problem-Solving Practice

Solve.

5 **TEMPERATURE** Mr. Evans recorded the average temperature for an hour and a half for one week. Use the data from the table to make a line graph. Between what times did the temperature decrease the most?

Average Time	6:00 A.M.	6:15 A.M.	6:30 A.M.	6:45 A.M.	7:00 A.M.	7:15 A.M.	7:30 A.M.
Temperature (°F)	8	11	13	9	14	12	9

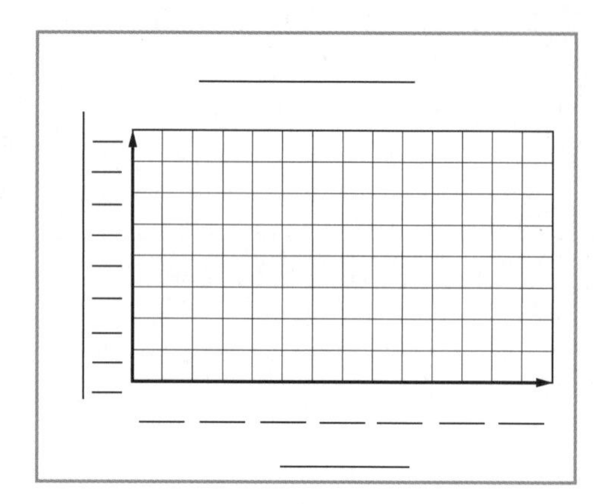

Check off each step.

______ **Understand: I underlined key words.**

______ **Plan: To solve the problem, I will** ________________________________.

______ **Solve: The answer is** ________________________________.

______ **Check: I checked my answer by** ________________________________

________________________________.

Skills, Concepts, and Problem Solving

Use the line graph to make a table of the data.

6

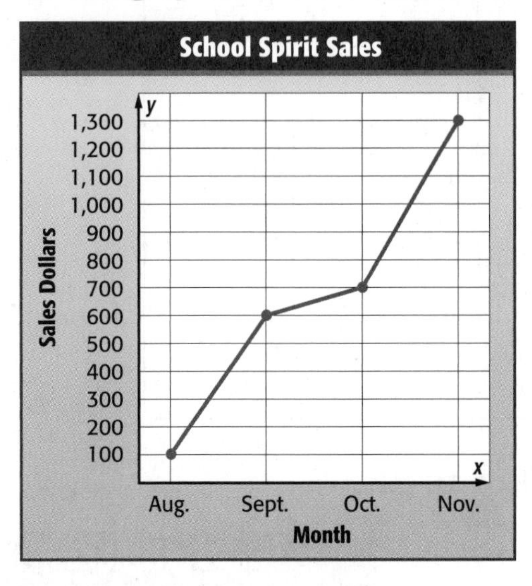

____

Month	Sales ($)

7

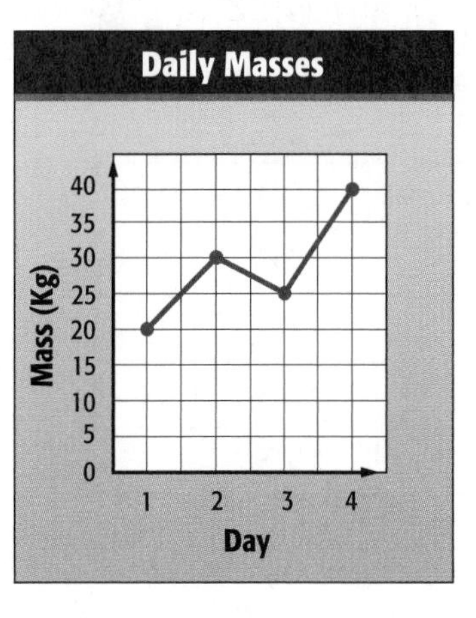

____

Day	Mass

8 The table below shows the annual gas prices between 2000 and 2006. Make a line graph of the data. Describe the change in data.

Year	Gas Price ($) per Gallon
2000	1.93
2001	1.85
2002	1.68
2003	1.99
2004	2.25
2005	2.24
2006	2.81

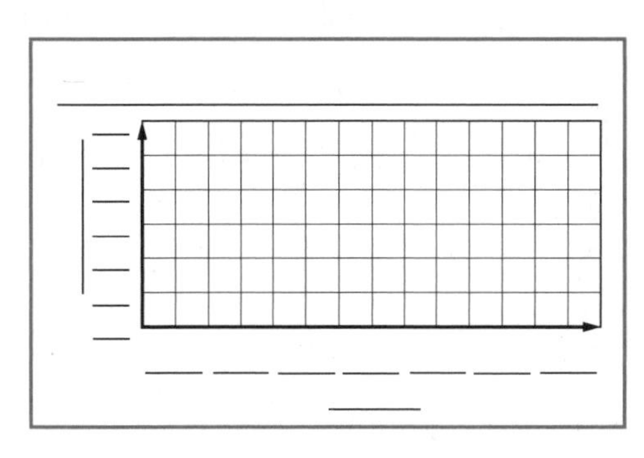

____

____

GO ON

9 The table below shows sales of laundry detergent during the spring and summer months. Make a line graph of the data. Describe the change in data.

Month	Laundry Detergent Sales ($)
March	75
April	100
May	125
June	80
July	95
August	175

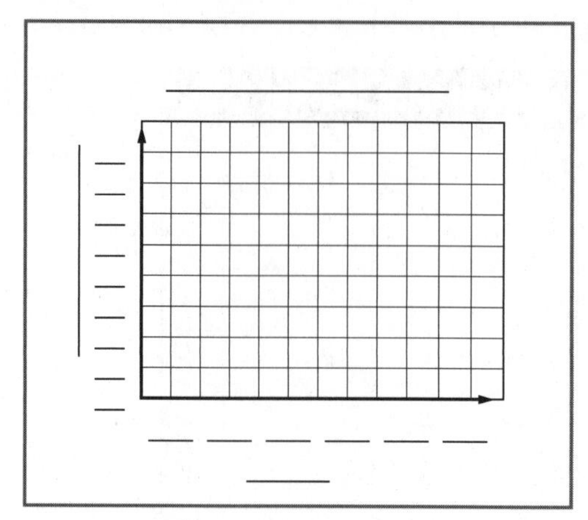

HEALTH **Use the following line graph to answer Exercises 10–12.**

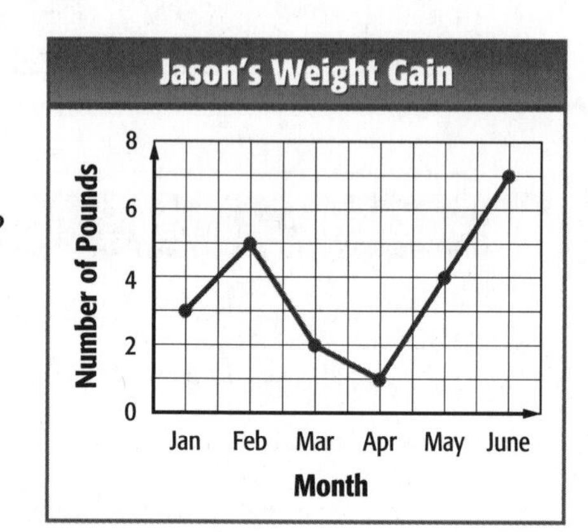

10 During which month did Jason gain the most weight?

____

11 How many pounds did Jason gain between April and June?

____

12 Between which two months did Jason's weight loss decrease the most?

____

Vocabulary Check **Write the vocabulary word that completes each sentence.**

13 A(n) ______________ shows how a set of data changes over a period of time.

14 The axis on which the categories or values are shown is the ______________.

15 The y-axis is also known as the ______________.

16 **Reflect** How do you determine your axes' titles when labeling a line graph?

____

____

____

STOP

Lesson 6-2

Frequency Tables and Histograms

KEY Concept

A **frequency** table displays data in equal **intervals**. In this table, the interval is 14 years.

Age	Frequency
0–14	45
15–29	29
30–44	37
45–59	22
60–74	18
75–89	7

The scale is 0 to 89.

Frequency is how often data occurs within that interval.

Data from a frequency table can be graphed in a **histogram**.

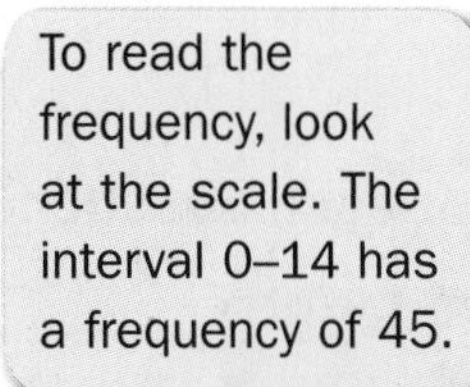

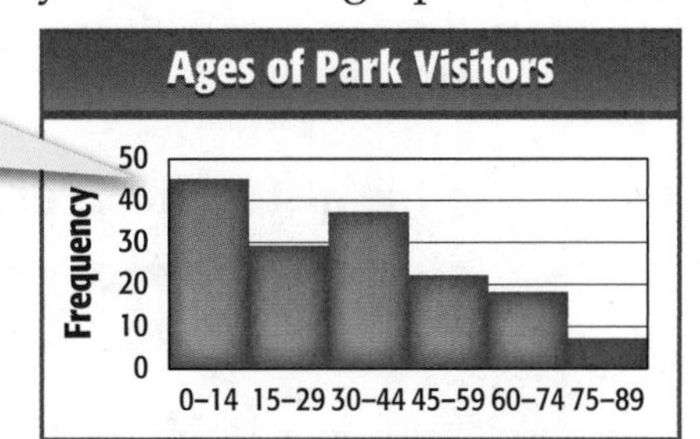

A histogram has bars with the same width that touch.

VOCABULARY

frequency
how often a piece of data in that interval occurs

frequency table
a chart that indicates the number of values in each interval

histogram
a graphical display that uses bars to display numerical data that have been organized into equal intervals

intervals
the difference between successive values on a scale

scale
the set of all possible values in a given measurement, including the least and greatest numbers in the set, separated by the intervals used

Example 1

Use the data from the frequency table to complete the histogram.

Minutes Spent Walking	0–4	5–9	10–14	15–19	20–24
Frequency	5	7	8	19	15

1. The first interval is 0–4 minutes.
   There is a frequency of 5 in the first interval.
2. Draw a bar up to the 5 on the vertical axis and color in the bar.
3. Repeat Step 2 for all intervals.

Bars are placed next to each other and are the same width.

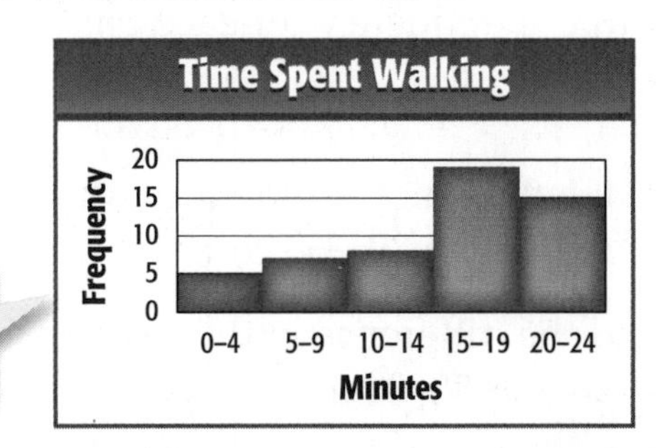

GO ON

YOUR TURN!

Use the data from the frequency table to complete the histogram.

Hours Exercising	0-1	2-3	4-5	6-7	8-9
Frequency	2	8	11	7	3

1. The first interval is ________. There is a frequency of ________ in the first interval.
2. Draw a bar up to the ________ on the vertical axis and color in the bar.
3. Repeat Step 2 for all intervals.

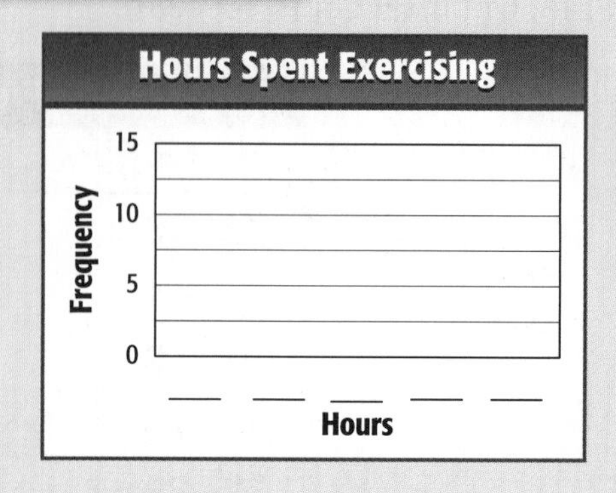

Example 2

Use the histogram to answer each question.

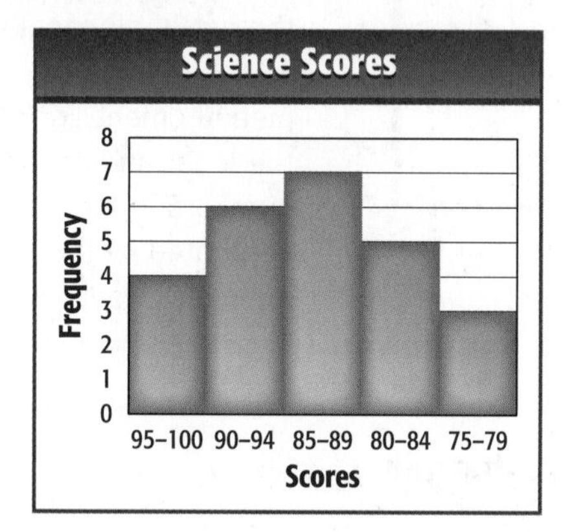

1. How many students' scores are shown in the histogram?

   Add the frequency for each bar.

   $4 + 6 + 7 + 5 + 3 = 25$

2. How many more students earned a 90–94 score than an 80–84 score?

   Find the difference in the frequencies for the 90–94 and 80–84 intervals.

   $6 - 5 = 1$

YOUR TURN!

Use the histogram to answer each question.

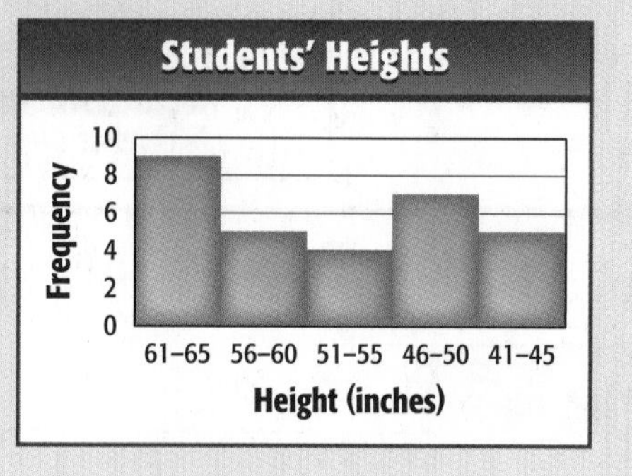

1. How many students' heights are shown in the histogram?

   ______ the ________ for each bar.

   9 + ____ + ____ + ____ + ____ = ____

2. How many more students are in the 46–50 inch interval than 56–60 inch interval?

   Find the differences in frequencies for the ________ and ________ intervals.

   ______ – ______ = ______

Guided Practice

Use the data from the frequency table to complete the histogram.

1

Temperature	Frequency
40–49	2
50–59	7
60–69	9
70–79	10
80–89	2

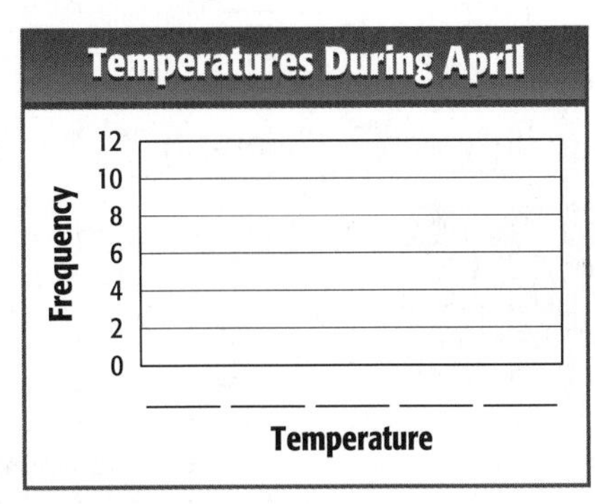

Step by Step Practice

2 Use the histogram to answer each question.

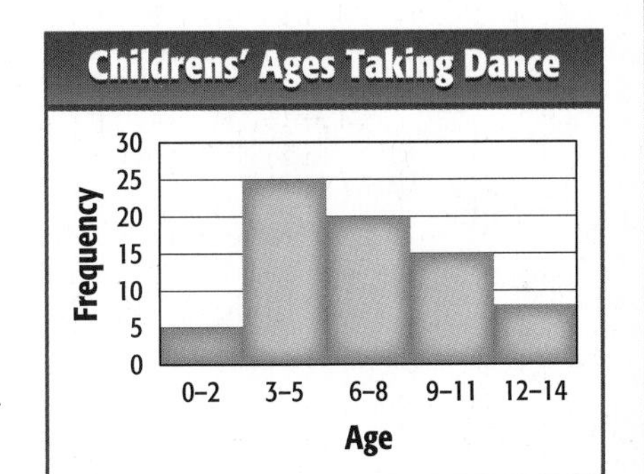

Step 1 How many dancers are represented in the histogram?

Add the ____________ for each bar.

_____ + _____ + _____ + _____ + _____ = _____

Step 2 How many more dancers are in the 3–5 year old interval than the 9–11 year old interval?

Find the differences in frequencies for the 3–5 and 9–11 intervals.

_____ − _____ = _____

_____ more dancers are in the 3–5 interval than the 9–11 interval.

Use the data from the frequency table and histogram in Exercise 1 to answer Exercises 3–5.

3 How many days was the temperature recorded? _____

4 How many more days were in the 70s than the 40s? _____

5 How many more days had a high temperature in the 60s than in the 50s? _____

Step by Step Problem-Solving Practice

Solve.

6 **SALES TAX** Use the data from the frequency table to complete the histogram. How many of the 50 states do not have a state sales tax rate between 2.1% and 8.0%?

Sales Tax (%)	2.1–3.0	3.1–4.0	4.1–5.0	5.1–6.0	6.1–7.0	7.1–8.0
Frequency (number of states)	1	7	11	16	9	1

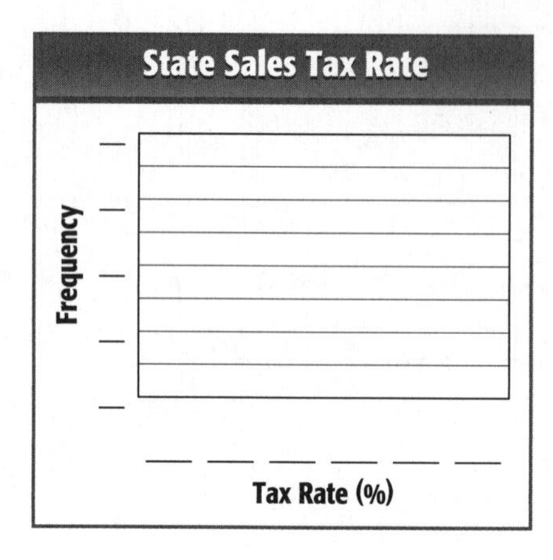

____

____

Check off each step.

_____ **Understand: I underlined key words.**

_____ **Plan: To solve the problem, I will** _______________.

_____ **Solve: The answer is** _______________.

_____ **Check: I checked my answer by** _______________.

Skills, Concepts, and Problem Solving

The histogram shows the number of tardy slips given to students at Weston Middle School during the first semester. Use this histogram to answer Exercises 7–9.

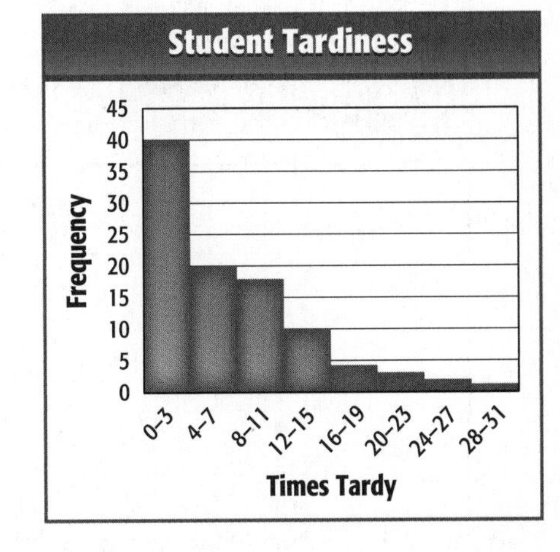

7 About how many students are represented in this graph?

____

8 How many tardy slips have most students received?

____

9 How many more students received 4–7 tardy slips than 12–15?

____

TECHNOLOGY

The results of a survey are shown in the histogram. Use the histogram to answer Exercises 10 and 11.

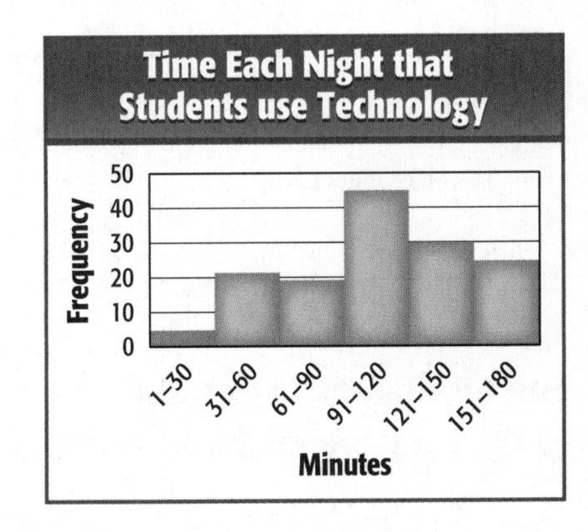

10 About how many students use technology one hour or less each evening?

____

11 About how many more students use technology 30 minutes or less compared to more than 150 minutes?

____

Vocabulary Check **Write the vocabulary word that completes each sentence.**

12 A(n) _______________ indicates the number of values in each interval.

13 _______________ are the difference between successive values on a scale.

14 Reflect Using the histogram for Exercises 10 and 11, can it be determined that the number of students that spend 30 minutes on a computer each night? Explain.

____

____

Chapter 6

Progress Check 1 (Lessons 6-1 and 6-2)

1 The data in the table shows the average bread prices from 2003 to 2008. Make a line graph of this data.

Year (Jan)	2003	2004	2005	2006	2007	2008
Bread Price ($) (U. S. City Average)	1.04	0.95	1.00	1.05	1.15	1.28

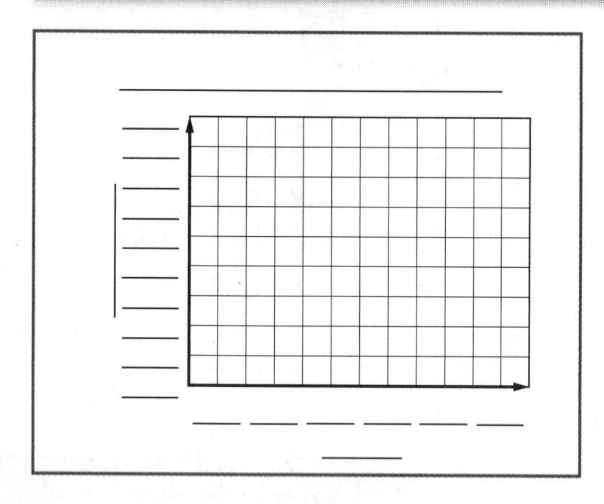

2 How many times did a decrease in bread price occur from one year to the next?

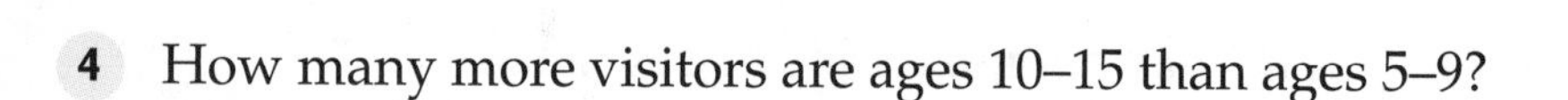

The histogram shows the number of visitors by age who were admitted to an amusement park. Use this histogram to answer Exercises 3–6.

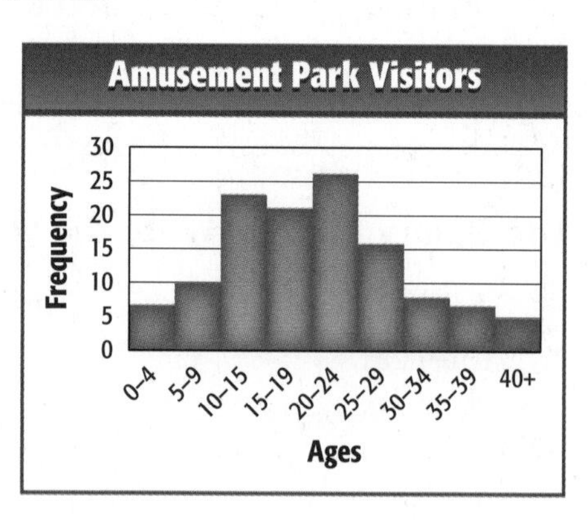

3 Which age interval is most represented?

4 How many more visitors are ages 10–15 than ages 5–9?

5 What group is least represented?

6 Between which two intervals shows the least decrease?

Lesson 6-3

Bar Graphs

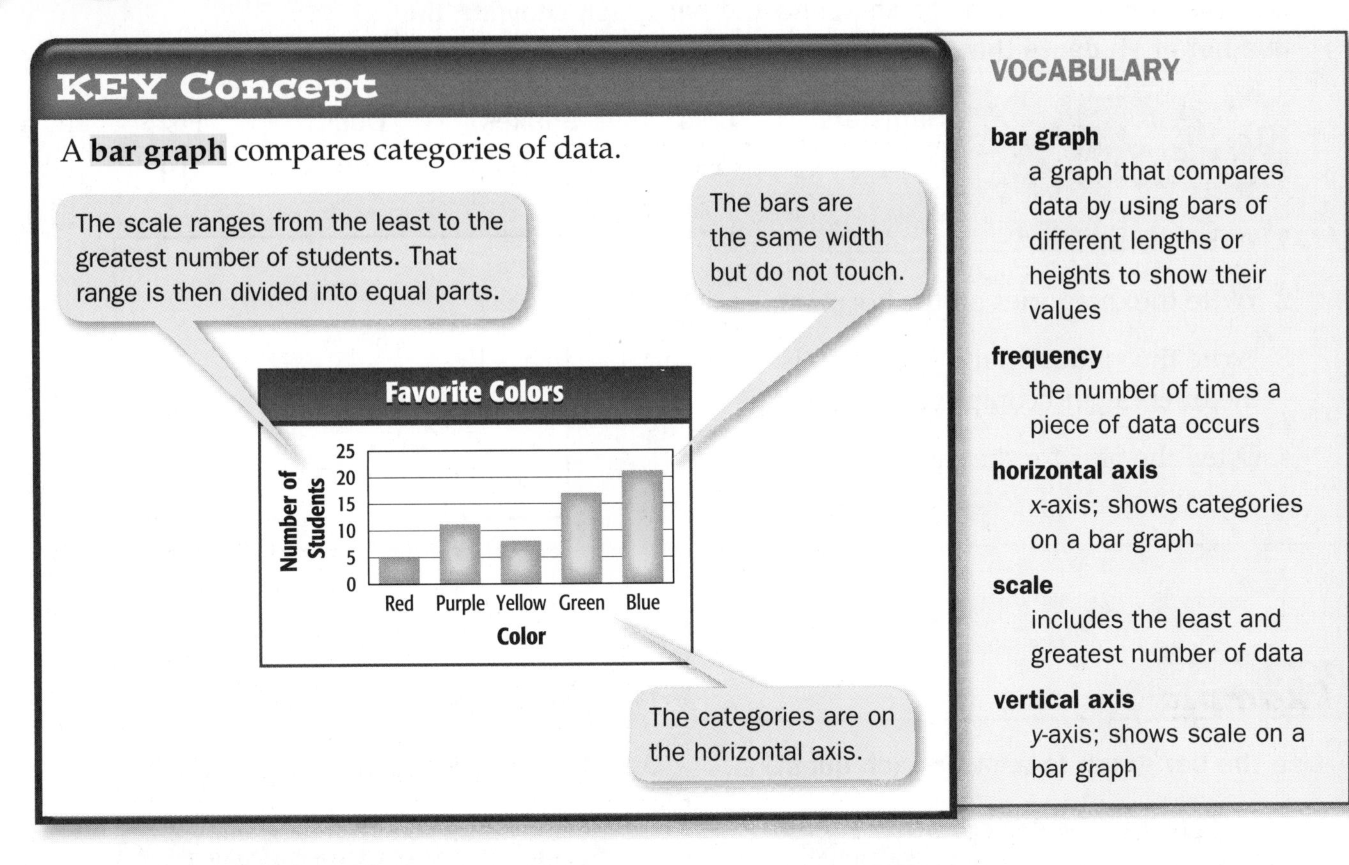

VOCABULARY

bar graph
a graph that compares data by using bars of different lengths or heights to show their values

frequency
the number of times a piece of data occurs

horizontal axis
x-axis; shows categories on a bar graph

scale
includes the least and greatest number of data

vertical axis
y-axis; shows scale on a bar graph

A bar graph is a good way to display data when a quick at-a-glance comparison is needed.

Example 1

Use the data in the table to complete the bar graph showing the number of students in afterschool activities.

Activities	Clubs	Sports	Lessons	Tutor	Homework
Number of Students	18	33	7	6	25

1. Write the title.

   Afterschool Activities

2. Write the categories on the horizontal axis.
3. Write the number values on the vertical axis. Choose the interval and complete the scale.
4. Draw the bars for the categories.

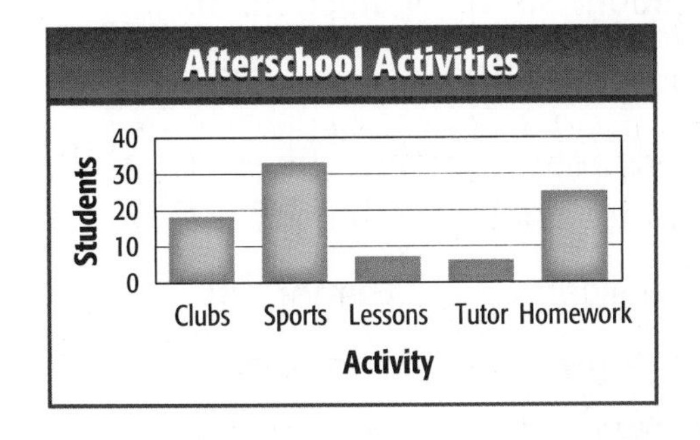

GO ON

YOUR TURN!

Use the data in the table to complete the bar graph showing the number of students that own certain pets.

Pets	Hamsters	Cats	Snakes	Dogs	Fish
Number of Pets	7	41	3	30	19

1. Write the title. ______________
2. Write the categories on the horizontal axis.
3. Write the values on the vertical axis. Choose the interval and complete the scale.
4. Draw the bars for the categories.

Example 2

Use the bar graph to answer each question.

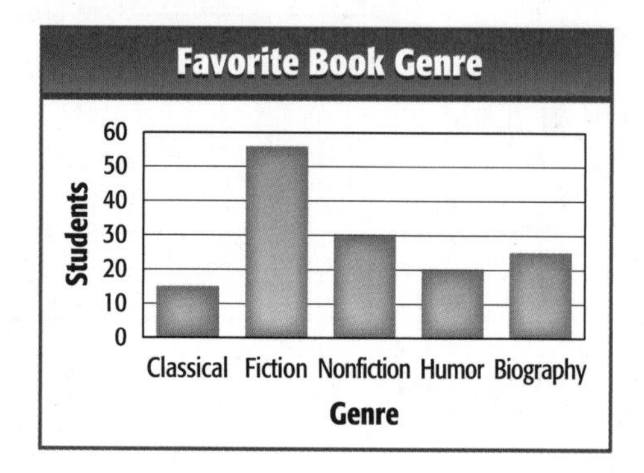

1. How many more students liked fiction books than nonfiction?

   Fiction: 55 **Nonfiction:** 30

   Find the difference. $55 - 30 = 25$

2. How many students liked classical and humor books combined?

   Classical: 15 **Humor:** 20

   Find the sum. $15 + 20 = 35$

YOUR TURN!

Use the bar graph to answer each question.

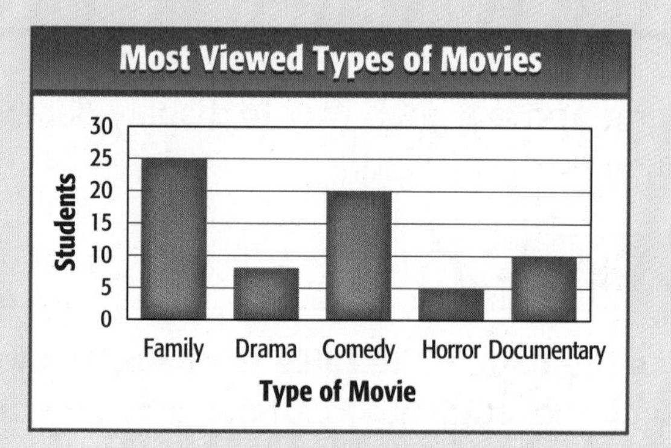

1. How many more students watch comedy than drama?

   Comedy: _____ Drama: _____

   Find the difference. _____ − _____ = _____

2. How many students view documentaries and family movies combined?

   Documentaries: _____ Family: _____

   Find the sum. _____ + _____ = _____

Guided Practice

1 Use the data to complete the bar graph.

Grade Level	Amount of Money Raised
9th	$458
10th	$620
11th	$197
12th	$381

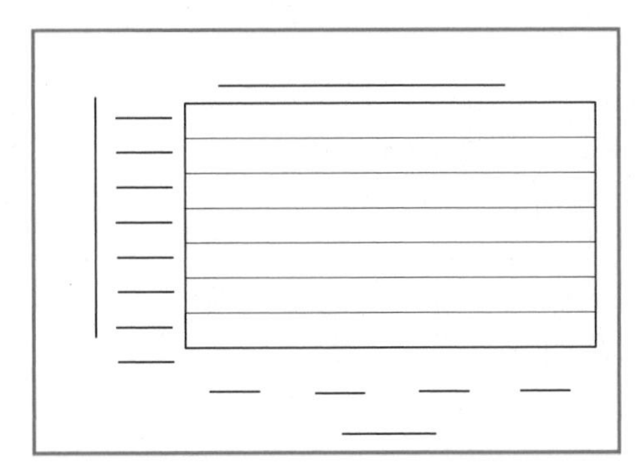

Step by Step Practice

2 Use the bar graph to answer each question.

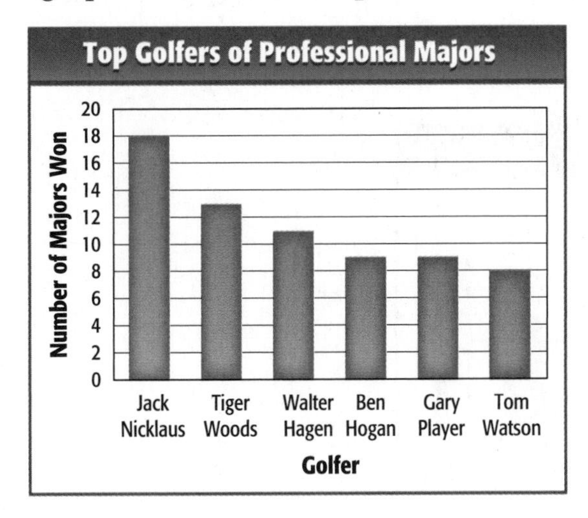

Step 1 How many more majors has Tiger Woods won than Ben Hogan?

Woods won _____ majors. Hogan won _____ majors.

Find the difference.

_____ − _____ = _____

Step 2 How many majors has Woods and Nicklaus won combined?

Nicklaus won _____ majors. Woods won _____ majors.

Find the sum.

_____ + _____ = _____

GO ON

Use the bar graph to answer Exercises 3–4.

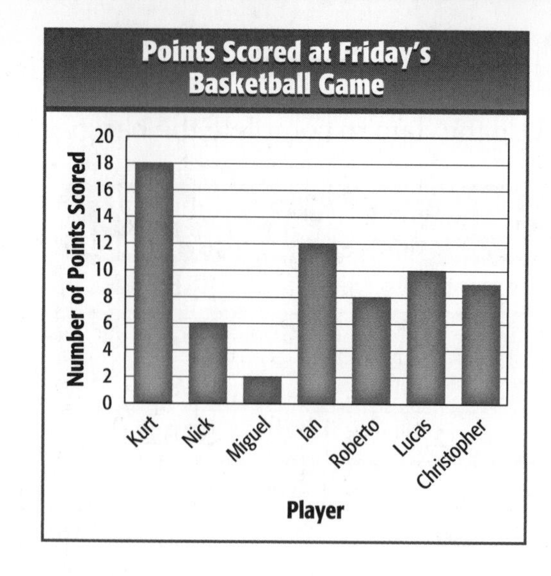

3. How many more points did Ian score than Miguel?

   _____ − _____ = _____ points

4. How many points did Kurt, Nick, and Roberto score combined?

   _____ + _____ + _____ = _____ points

Step by Step Problem-Solving Practice

Solve.

5. **SCHOOL** Use the bar graph to find how many more students prefer art or music than English.

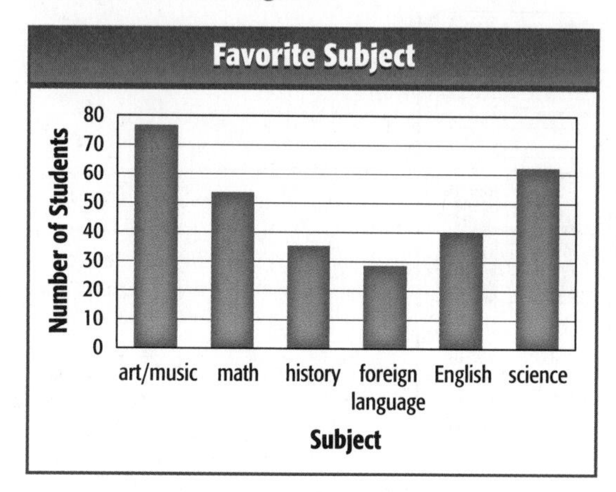

Check off each step.

_____ **Understand: I underlined key words.**

_____ **Plan: To solve the problem, I will** _______________.

_____ **Solve: The answer is** _______________.

_____ **Check: I checked my answer by** _______________

_______________.

Skills, Concepts, and Problem Solving

6 Use the data in the table to create a bar graph that shows the enrollment of students each year at Swanson High School.

School Year	Number of Students
2001	452
2002	489
2003	504
2004	477
2005	460
2006	482
2007	495

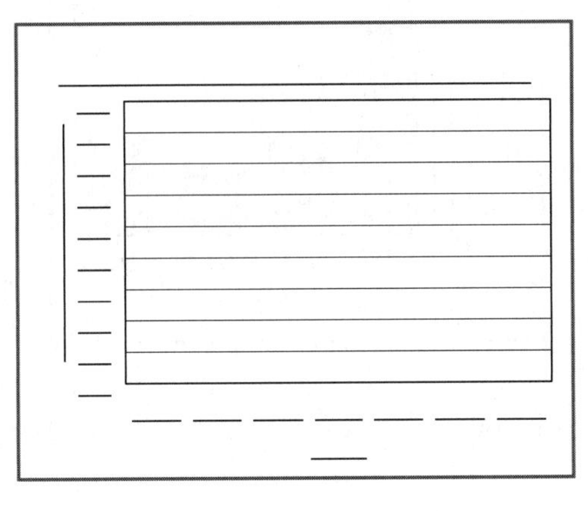

Use the bar graph to answer Exercises 7–12.

7 Which year had the highest enrollment at Swanson High?

____

8 How many students enrolled in Swanson High in 2005?

____

9 How many more students enrolled in 2006 than 2001?

____

10 Which year had the lowest enrollment at Swanson High?

____

11 How many more students enrolled in 2003 than 2005?

____

12 How many students enrolled in Swanson High between 2002 and 2006?

____

GO ON

MONEY **Use the bar graph to answer Exercises 13–15.**

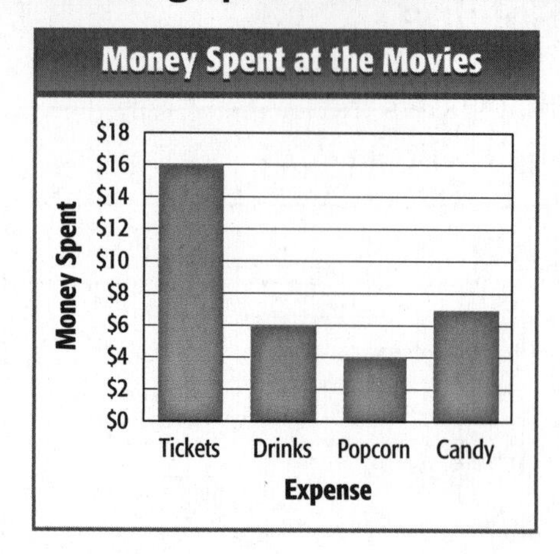

13 How much money did Dennis and Camila spend at the movies?

____

14 How much more money was spent on tickets than on popcorn?

____

15 How much money was spent on candy, popcorn, and drinks combined?

____

Vocabulary Check **Write the vocabulary word that completes each sentence.**

16 A(n) ______________ compares data by using bars of different lengths or heights to show their values.

17 The ______________ is the number of times a piece of data occurs.

18 The scale on a bar graph is located on the ______________.

19 The categories on a bar graph are located on the

______________.

20 **Reflect** Can a bar graph be used to display any data set? Explain.

____

____

____

STOP

Lesson 6-4 Circle Graphs

KEY Concept

A **circle graph** shows how parts of data are related to the whole set of data.

Favorite Lunch Foods

The parts or sectors add up to the whole, or 100%.

The percents in a circle graph add up to 100%.

19% + 37% + 13% + 10% + 21% = 100%

You can analyze data in a circle graph by multiplying the percent by the whole. For example, if 500 students were surveyed, 21% • 500 = 105 students preferred chicken.

VOCABULARY

central angle
an angle whose vertex is the center of the circle

data
pieces of information, which are often numerical

sector
pie-shaped sections on a circle graph

Example 1

Use the data in the graph to answer each question.

Device Used Most

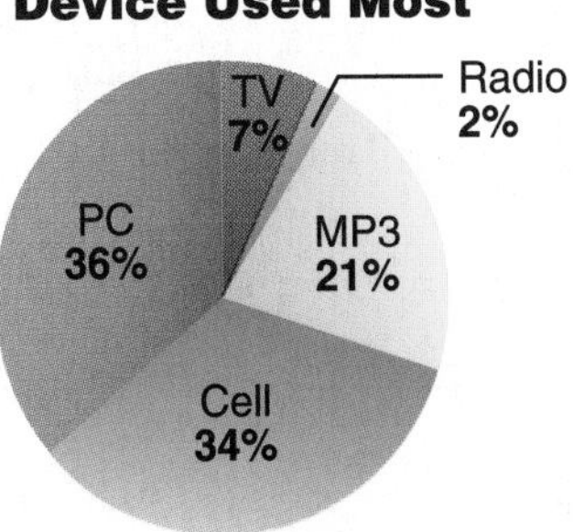

1. If 400 students were surveyed, how many students said they used their cell phones most?

   400 • 34% = 400 • 0.34 = 136
   136 students said their cell phones.

2. Which device was chosen the least?

   The smallest sector is for radio. So, radio was chosen the least.

YOUR TURN!

Use the data in the graph to answer each question.

Favorite Exercise

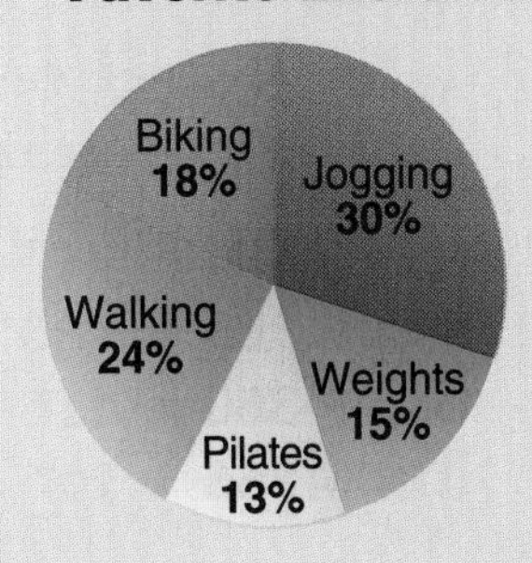

1. If 250 students were surveyed, how many students said they liked biking the most?

   250 • ______ = 250 • ______ = ______

   ______ students liked biking the most.

2. Which activity was chosen the most?

   The largest sector is for __________. So, __________ was chosen the most.

GO ON

Example 2

The table shows how students at Sherman Woods High School get to school. Create a circle graph using the data in the table.

Way to School	
Walk	52
Bus	72
Car	64
Other	12

1. Find the total number of students surveyed.

   $52 + 72 + 64 + 12 = 200$

   200 is the whole.

2. Divide each part by the whole to find the percent for each sector.

   $\frac{52}{200} = 0.26 = 26\%$

   $\frac{72}{200} = 0.36 = 36\%$

   $\frac{64}{200} = 0.32 = 32\%$

   $\frac{12}{200} = 0.06 = 6\%$

   The total of the percents should equal 100%.

3. To find the degrees of each sector, multiply each percent decimal by 360.

   $0.26 \cdot 360 \approx 94°$

   $0.36 \cdot 360 \approx 130°$

   $0.32 \cdot 360 \approx 115°$

   $0.06 \cdot 360 \approx 21°$

   The degrees of the sectors should total 360.

4. Use a protractor to draw each sector. Label each sector with the category and percent.

Way to get to School

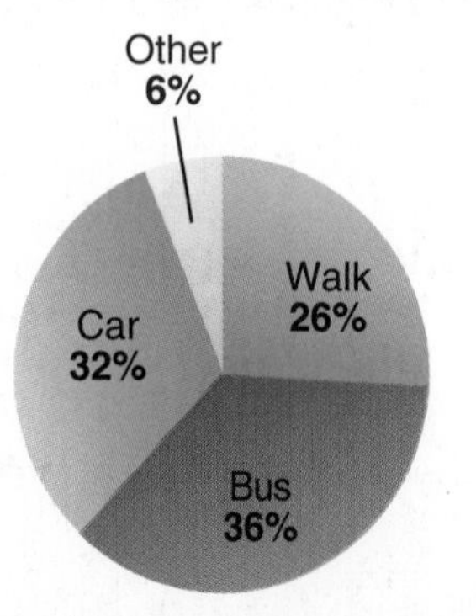

YOUR TURN!

The table shows students' favorite fruits. Create a circle graph using the data in the table.

Favorite Fruit	
Banana	33
Orange	48
Apple	39
Grapes	30

1. Find the total number of students surveyed.

   $33 + 48 + 39 + 30 =$ ________

   ________ is the ________.

2. Divide each part by the whole to find the percent for each sector.

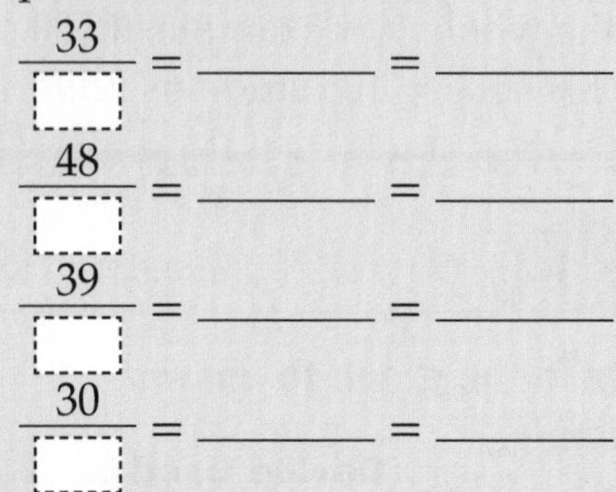

   $\frac{33}{\square} =$ ________ $=$ ________

   $\frac{48}{\square} =$ ________ $=$ ________

   $\frac{39}{\square} =$ ________ $=$ ________

   $\frac{30}{\square} =$ ________ $=$ ________

3. To find the degrees of each sector, multiply each percent decimal by 360.

   ________ $\cdot 360 \approx$ ________

   ________ $\cdot 360 \approx$ ________

   ________ $\cdot 360 \approx$ ________

   ________ $\cdot 360 \approx$ ________

4. Use a protractor to draw each sector. Label each sector with the category and percent.

____

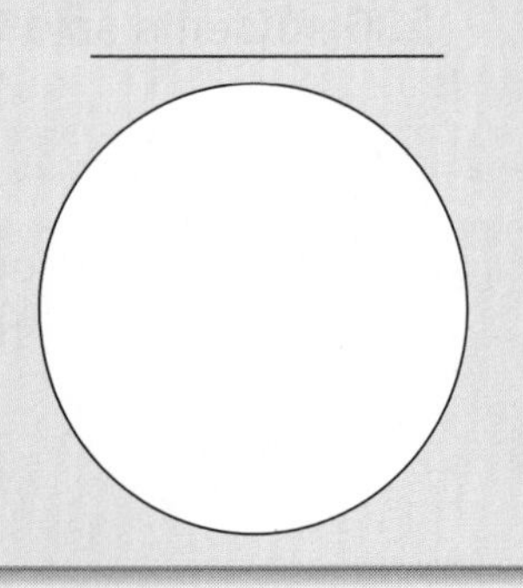

Guided Practice

Use the data in the graph at the right to answer Exercises 1–4.

Color of Car

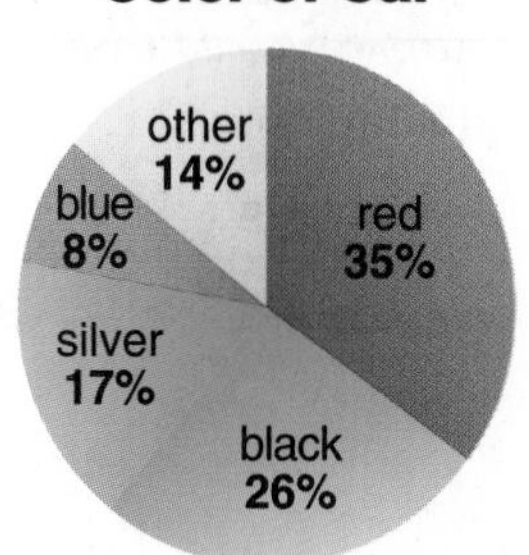

1. If 300 students were surveyed, how many students said they drive a silver car?

   300 • ______% = 300 • ______ = ______

   ______ students drive a silver car.

2. If 500 students were surveyed, how many students said they drive a black car?

   ______ • ______% = ______ • ______ = ______

   ______ students drive a black car.

3. Which car color is driven least? __________

4. Which car color is driven most? __________

Step by Step Practice

5. The table shows whether students in the tenth grade drive to school. Create a circle graph using the data.

Tenth Grade Students That Drive to School	
yes	64
no	136

Step 1 Find the total number of students surveyed.

______ + ______ = ______

______ is the whole.

Step 2 Divide each part by the whole to find the percent for each sector.

$\frac{64}{\square}$ = ______ • 100 = ______ $\qquad$ $\frac{136}{\square}$ = ______ • 100 = ______

Step 3 To find the degrees of each sector, multiply each percent decimal by 360.

______ • 360 ≈ ______ $\qquad$ ______ • 360 ≈ ______

Step 4 Use a protractor to draw each sector. Label each sector with the category and percent.

____

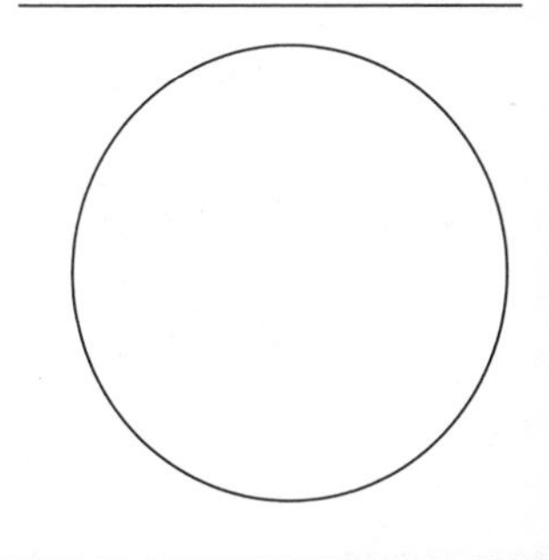

6 Create a circle graph using the data.

Volunteers for School Play	
singers	26
dancers	35
extras	31
stage crew	16

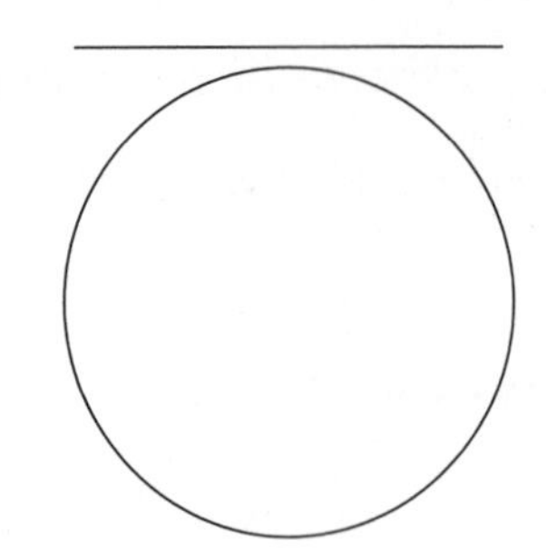

_____ + _____ + _____ + _____ = _____

singers

$\frac{26}{\square} \approx$ _____ = _____

_____ · 360 ≈ _____

dancers

$\frac{35}{\square} \approx$ _____ = _____

_____ · 360 ≈ _____

extras

$\frac{31}{\square} \approx$ _____ = _____

_____ · 360 ≈ _____

stage crew

$\frac{16}{\square} \approx$ _____ = _____

_____ · 360 ≈ _____

Step by Step Problem-Solving Practice

Solve.

7 SUMMER Six hundred forty students were surveyed about what they missed most about school during summer break. How many said they miss sporting events during summer break?

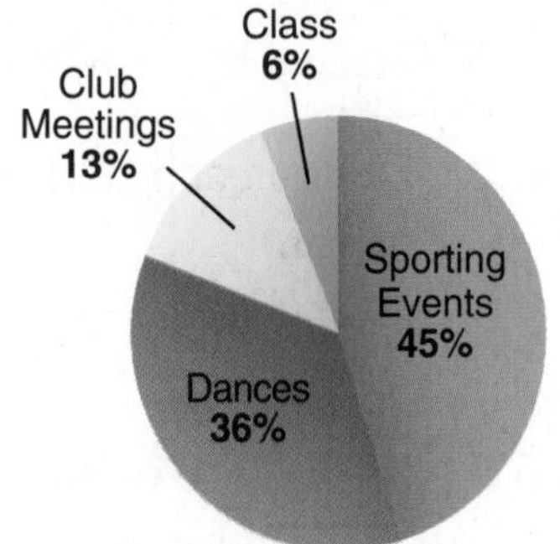

_____ · _____ = _____

Check off each step.

_____ Understand: I underlined key words.

_____ Plan: To solve the problem, I will _____.

_____ Solve: The answer is _____.

_____ Check: I checked my answer by _____

_____.

Skills, Concepts, and Problem Solving

Use the data in the table to answer Exercises 8–11.

Ways Students Earn Money	
cutting grass	18
babysitting	30
allowance	12
work at a store	48
other	12

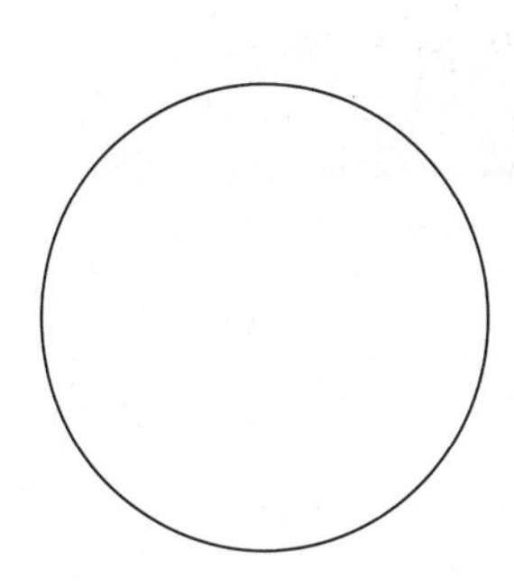

8. Make a circle graph using the data in the table.

9. How many total students were surveyed? ______

10. What percentage of students cut grass? ______

11. What percentage of students babysit? ______

HEALTH Use the circle graph with Hally's daily calorie intake to answer Exercises 12–14.

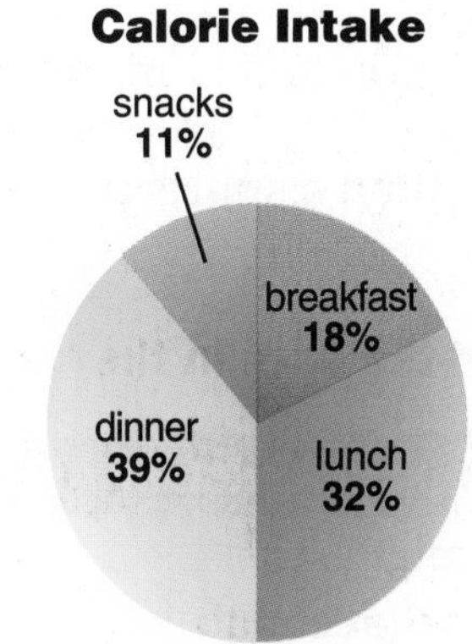

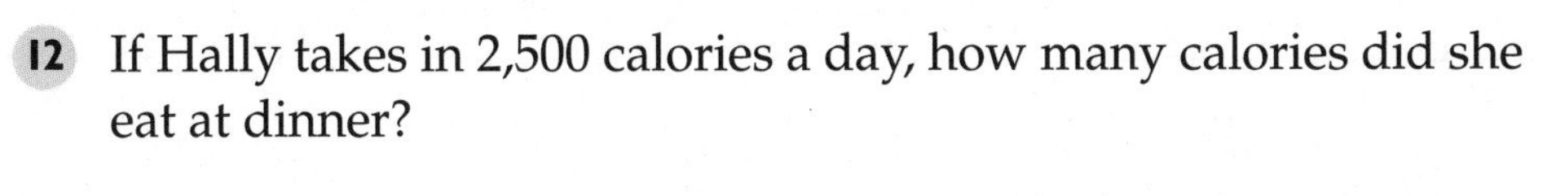

12. If Hally takes in 2,500 calories a day, how many calories did she eat at dinner?

____

13. How many more calories did Hally eat at lunch than at breakfast?

____

14. How many more calories did Hally eat at dinner than breakfast and snacks combined?

____

Vocabulary Check Write the vocabulary word that completes each sentence.

15. The ______________ is the angle whose vertex is the center of the circle.

16. **Reflect** When might you use a circle graph to display data?

____

Chapter 6

Progress Check 2 (Lessons 6-3 and 6-4)

The data in the table shows the type of drink chosen by Lake High School students at lunch time. Use the data for Exercises 1–4.

Drink	Number of Students
orange juice	139
lemonade	132
chocolate milk	128
water	145
white milk	130
sports drink	86

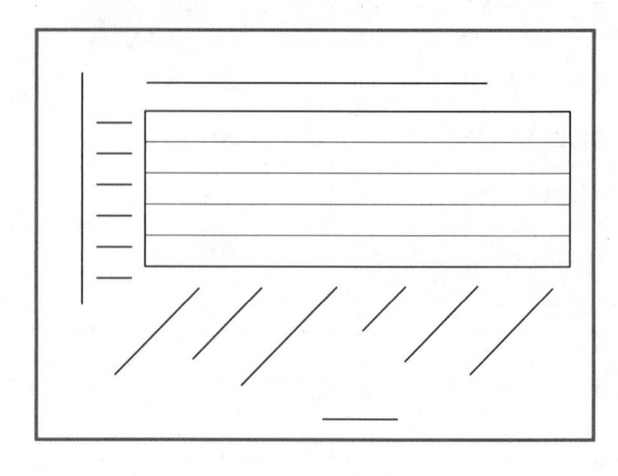

1. Make a bar graph to display the data.
2. Which drink was chosen the least? ______________
3. How many students chose lemonade? ______________
4. How many more students chose white milk than a sports drink? ______________

Use the data in the table to answer Exercises 5–8.

Extracurricular Activities	
foreign language club	12
sports	27
art	18
service	24
other	9

____

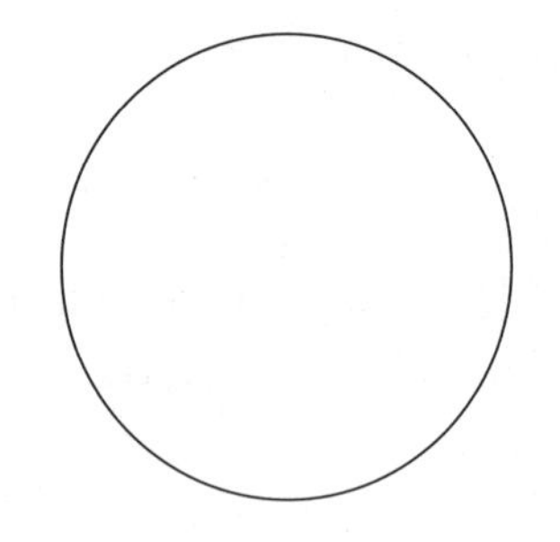

5. How many total students were surveyed? ______________
6. What percentage of students work on art? ______________
7. What percentage of students play sports? ______________
8. Make a circle graph using the data in the table.

Lesson 6-5

Mean, Median, and Mode

KEY Concept

Measures of central tendency describe a data set. The most common measures are **mean**, **median**, and **mode**. Consider the data set 4, 10, 10, 8, 3.

The mean is the sum of all data in the set divided by the number of elements in the set.

$$\text{mean} = \frac{4 + 10 + 10 + 8 + 3}{5} = \frac{35}{5} = 7$$

The mean is 7.

The median is the data element in the middle or the mean of the two middle data elements of data listed in numerical order.

3, 4, [8], 10, 10

The median is 8.

The mode is the data element that occurs most often.

4, (10), (10), 8, 3

The mode is 10.

VOCABULARY

mean
the sum of the numbers in a set of data divided by the number of items in the data set

measures of central tendency
numbers or pieces of data that can represent the whole set of data

median
the middle number in a set of data when the data are arranged in numerical order; if the data set has an even number, the median is the mean of the two middle numbers

mode
the item(s) that appear most often in a set of data

Example 1

Find the mean, median, and mode for the given data set.

23, 11, 15, 20, 9, 11, 23

1. Write the data in order from least to greatest. 9, 11, 11, 15, 20, 23, 23
2. Find the mean. Add the numbers. Divide by 7.

   $9 + 11 + 11 + 15 + 20 + 23 + 23 = 112$
   $112 \div 7 = 16$

3. Find the median. The middle number is 15.
4. Find the mode. 11 and 23 appear in the list twice.

YOUR TURN!

Find the mean, median, and mode for the given data set.

51, 15, 15, 23, 36

1. Write the data in order from least to greatest. ________________
2. Find the mean.Add the numbers. Divide by ________.

   ____

   ____

3. Find the median. The middle number is ________.
4. Find the mode. ________ appears in the list ________.

Example 2

Use the data in the table to answer the questions.

Basketball Scores		
Game	Liz	Tina
1	11	23
2	27	19
3	9	7
4	6	12
5	14	19

1. Which player has the greater mean score?

   Find the mean for each player.

   Liz:
   $11 + 27 + 9 + 6 + 14 = 67$
   $67 \div 5 = 13.4$

   Tina:
   $23 + 19 + 7 + 12 + 19 = 80$
   $80 \div 5 = 16$
   Tina has the greater mean.

2. Who has the lower median score?

   Liz: 6, 9, 11, 14, 27
   The median is 11.

   Tina: 7, 12, 19, 19, 23
   The median is 19.

   Liz **has the lower median score.**

YOUR TURN!

Use the data in the table to answer the questions.

Students' Scores		
Student	Math Class	Science Class
1	89	92
2	97	78
3	67	90
4	73	69
5	88	82

1. Which class has the greater mean score?

   Find the mean for each class.

   Math:

   ____

   ____

   Science:

   ____

   ____

   __________ has the greater mean score.

2. Which class has the greater mode score?

   ____

Guided Practice

1 Find the mean, median, and mode.
9, 11, 13, 6, 7, 4, 6

List the data from least to greatest. ______________________

The mean is __________ ÷ __________ = __________.

The median is __________.

The mode is __________.

Step by Step Practice

2 Use the data in the table to answer the questions.

Temperatures		
Day	**Week 1**	**Week 2**
Sun.	65°	66°
Mon.	58°	73°
Tues.	54°	80°
Wed.	63°	76°
Thurs.	76°	71°
Fri.	84°	67°
Sat.	71°	69°

Step 1 Which week has the lower median temperature?

Week 1: ______________________

Week 2: ______________________

__________ has the lower median score.

Step 2 Which week has the greater mode temperature?

____

Step 3 Does Week 1 or 2 have the greater mean temperature?

Week 1: ______________________

____

Week 2: ______________________

____

__________ has the greater mean.

Find the mean, median, and mode to answer Exercises 3–5.

Weight in Pounds						
pumpkins	13	20	18	15	24	28
watermelons	21	17	12	16	22	21

3 Do pumpkins or watermelons have the greater mean weight?

pumpkins:

_____ + _____ + _____ + _____ + _____ + _____ = _____

_____ ÷ _____ = _____

watermelons:

_____ + _____ + _____ + _____ + _____ + _____ = _____

_____ ÷ _____ = _____

_______________ have the greater mean weight.

4 Which fruit has the lower median weight?

pumpkins:

____

The median is _____.

watermelons:

____

The median is _____.

_______________ have the same median weight.

5 Which fruit has the greater mode?

pumpkins: _______________

watermelons: _____

_______________ have the greater mode.

Step by Step Problem-Solving Practice

Solve.

6. **WORK** Kenji needs to work an average of 8 hours per day. How many hours does Kenji need to work Friday in order to have a mean of 8 hours a day for the week?

Kenji's Daily Hours Worked					
Day	**Monday**	**Tuesday**	**Wednesday**	**Thursday**	**Friday**
Hours	8	6.5	7	9	?

Find the total hours worked so far.

______ + ______ + ______ + ______ = ______

Find the total hours for the week if the mean is 8.

______ hours • ______ days = ______ hours

Subtract the hours worked from the week total.

____

Check your answer.

____

Check off each step.

______ **Understand: I underlined key words.**

______ **Plan: To solve the problem, I will** ______________________________.

______ **Solve: The answer is** ______________________________.

______ **Check: I checked my answer by** ______________________________

______________________________.

GO ON

Skills, Concepts, and Problem Solving

Find the mean, median, and mode for each data set. Round to the nearest tenth if needed.

Data Set	Mean	Median	Mode
20, 26, 18, 21, 23, 20	7 ______	8 ______	9 ______
8, 5, 6, 2, 9, 1, 8, 2	10 ______	11 ______	12 ______
56, 62, 48, 37, 50	13 ______	14 ______	15 ______

Solve. Write in simplest form.

16 **MEDIA** Maya wants to buy a magazine in the grocery checkout lane. The prices of the magazines are \$3.95, \$2.50, \$4.00, \$2.99, \$1.75. What is the mean price of the magazines? ______

17 **EXERCISE** Adina wants to exercise daily. She has recorded her exercising times over the past two weeks in a table. What is her mode time? ______

Adina's Exercise Log						
Sunday	Monday	Tuesday	Wednesday	Thursday	Friday	Saturday
45 min	60 min	30 min	50 min	60 min	25 min	0 min
45 min	40 min	0 min	15 min	30 min	0 min	55 min

Vocabulary Check **Write the vocabulary word that completes each sentence.**

18 The ______ is the middle number when the data is placed in order from least to greatest.

19 ______ is another word for average.

20 The number or numbers that occur most often is the ______.

21 **Reflect** Is the mode a good representation of the data in Exercise 17?

______

______

______

STOP

Lesson 6-6

Count Outcomes

KEY Concept

The **outcome** of an experiment is the result. To show all the different outcomes use a **tree diagram**.

If you toss a coin, the possible outcomes are heads or tails. If you toss a coin two times, there are 4 possible outcomes: HH, HT, TH, and TT.

The tree diagram connects the choices for the first toss and the second toss.

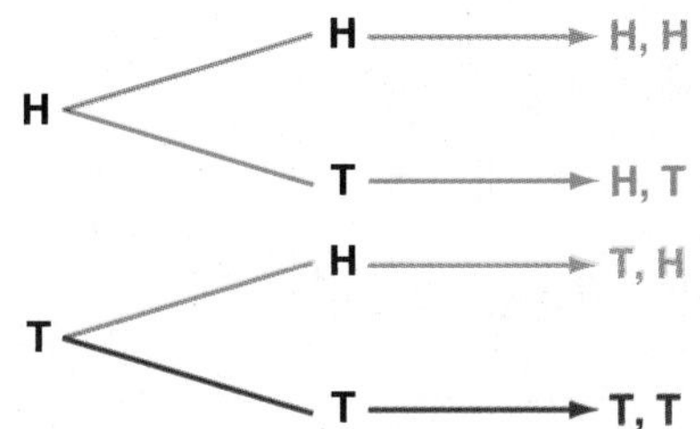

To find the total number of outcomes without a diagram, multiply using the **Fundamental Counting Principle**.

number of outcomes of first event (heads or tails)	•	number of outcomes of second event (heads or tails)	=	the total number of possible outcomes
2	•	2	=	4

VOCABULARY

event
a type of outcome

Fundamental Counting Principle
uses multiplication of the number of ways each event in an experiment can occur to find the number of total possible outcomes in a sample space

outcomes
the possible results of a probability event

tree diagram
a diagram used to show the total number of possible outcomes in a probability experiment

A tree diagram and the counting principle can be used for more than two events.

Example 1

Draw a tree diagram to find the outcomes if the spinner is spun twice.

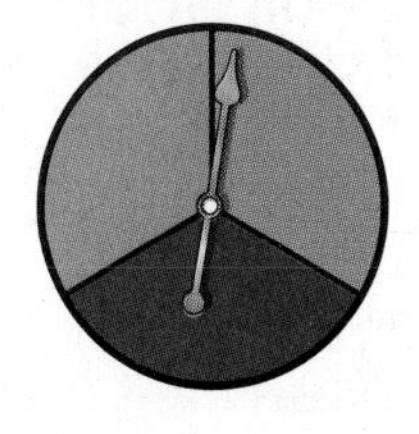

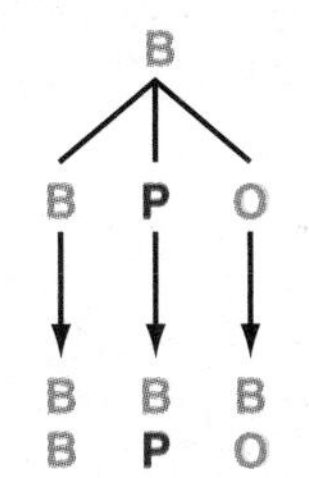

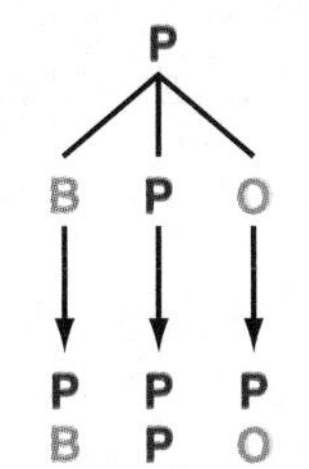

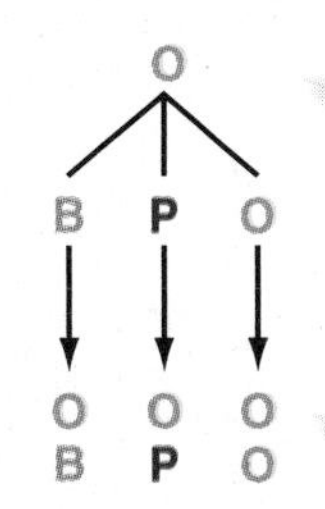

There are 3 possible outcomes for the first spin.

There are 3 possible outcomes for the second spin. Pair each with the first spin.

1. Use B for blue, P for purple, and O for orange.
2. List each outcome for the first spin.
3. Pair each outcome for the second spin with the first spin.
4. There are 9 total possible outcomes.

BB, BP, BO, PB, PP, PO, OB, OP, OO

GO ON

YOUR TURN!

Draw a tree diagram to find the outcomes if the spinner is spun twice.

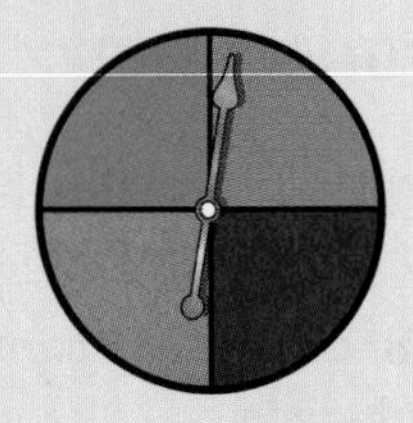

B P G O

1. Use B for blue, P for purple, G for green, O for orange.
2. List each outcome for the first spin.
3. Pair each outcome for the second spin with the first spin.
4. There are ______ total possible outcomes.

____

Example 2

Use the Fundamental Counting Principle to find the number of possible outcomes when one 6-sided number cube and one 2-sided coin are each tossed once.

1. For the number cube, there are 6 possible outcomes.
2. For the number coin, there are 2 possible outcomes.
3. Multiply the number of outcomes for each event.

   6 • 2 = 12

   There are 12 possible outcomes.

YOUR TURN!

Use the Fundamental Counting Principle to find the number of possible outcomes when one 6-sided number cube is tossed twice.

1. For the first number cube, there are

   ______ possible outcomes.
2. For the second number cube, there are

   ______ possible outcomes.
3. Multiply the number of outcomes for each event.

   ______ = ______

   There are ______ possible outcomes.

Guided Practice

Draw a tree diagram to find the outcomes.

1. Dion has to drive along two streets to get from Newport to Blackburn. The first street he can choose will be Miami Street (M), Second Street (S), or Windham (W). Then he can take Hilltop Road (H) or Victory Pike (V). How many different combinations of two streets can he choose?

M

S ________________________

W

Step by Step Practice

2. Use the Fundamental Counting Principle to find the number of possible outcomes when the following spinner is spun twice.

Step 1 For the first spin, there are ______ possible outcomes.

Step 2 For the second spin, there are ______ possible outcomes.

Step 3 Multiply the number of outcomes for each event.

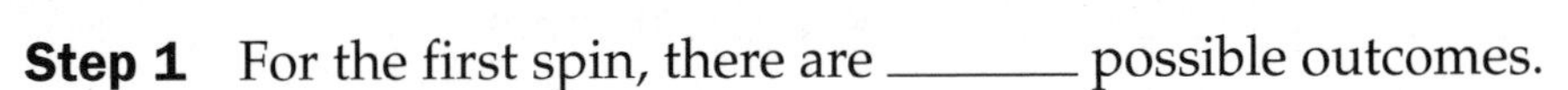

______ • ______ = ______.

There are ______ possible outcomes.

Use the Fundamental Counting Principle to find the number of possible outcomes for each situation.

3. Nicole is making a sandwich. She can have wheat or white bread. Her choices of meats are turkey, ham, or roast beef. She can have American or Swiss cheese.

number of choices of bread: ______

number of choices of meat: ______

number of choices of cheese: ______

______ • ______ • ______ = ____________

GO ON

4 Maria is choosing her work uniform. She can wear tan shorts or tan pants. Her four shirt choices are blue short-sleeved, blue long-sleeved, white short-sleeved, or white long-sleeved.

number of choices of pants: ______

number of choices of shirts: ______

______ • ______ = __________

Step by Step Problem-Solving Practice

Solve.

5 **PASSWORD** Byron needs to come up with a 4-digit password. The characters can only be letters or single-digit numbers (including 0). They can be used more than once. How many possible passwords can he make?

Find the total number of possible characters.

____________ + ____________ = ____________

Use the Fundamental Counting Principle.

______ • ______ • ______ • ______ = __________ possible outcomes

Check off each step.

______ **Understand: I underlined key words.**

______ **Plan: To solve the problem, I will** ______________________________.

______ **Solve: The answer is** ______________________________.

______ **Check: I checked my answer by** ______________________________.

Skills, Concepts, and Problem Solving

Draw a tree diagram to find the outcomes.

6 ICE CREAM The Sweet Treats Shoppe offers single-scoop ice cream in chocolate, vanilla, or strawberry, and two types of cones, regular or sugar. How many different combinations of the ice cream and cone are offered?

____

7 CLOTHES Adina can buy a school T-shirt with either short or long sleeves in small, medium, or large. How many different combinations can Adina choose from?

____

Use the Fundamental Counting Principle to find the number of possible outcomes for each situation.

8 Choose a vowel from the word *thought* and then choose a consonant from the word *front*.

____

9 Choose between the numbers 1, 2, 3, 4, and 5 and then choose between the colors red, blue, and yellow.

____

10 Choose between a quarter, dime, nickel, and penny and then choose a card from a deck of cards.

____

GO ON

RESTAURANT **Use the following menu to answer Exercises 11–13.**

Sunset Beach Grill

Fish	Cooking Method	Side Item
Halibut	Blackened	Fresh Vegetables
Salmon	Grilled	Baked Potato
Sea bass	Baked	Mashed Potatoes
Tuna	Fried	

11 How many outcomes are possible using Sunset Beach Grill's menu?

____

12 If Sunset Beach Grill ran out of salmon for the evening, how many outcomes are possible?

____

13 If Sunset Beach Grill added mackerel and cod to the types of fish available and french fries to the side item list, how many outcomes are possible?

____

Vocabulary Check **Write the vocabulary word that completes each sentence.**

14 The possible results of a probability event are ____________.

15 A type of outcome is known as a(n) ____________.

16 A(n) ____________ shows all possible outcomes.

17 **Reflect** When would you use the Fundamental Counting Principle instead of a tree diagram?

____

____

____

____

STOP

Lesson 6-7 Probability

KEY Concept

Probability is represented by numbers that tell how likely an **event** is to happen.

A probability of 0 means the event is impossible.

A probability of 0.5 means the event is as equally likely to happen as not to happen.

A probability of 1 means the event is certain.

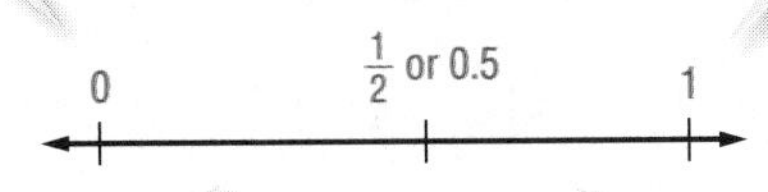

If a probability is between 0 and 0.5, the event is unlikely.

If a probability is between 0.5 and 1, the event is likely.

To calculate probability, divide the number of ways an event can occur, by the total number of possible **outcomes**.

$$P(\text{event}) = \frac{\text{favorable outcomes}}{\text{number of possible outcomes}}$$

VOCABULARY

event
a type of outcome

probability
the chance that an event will happen; the ratio of the number of ways an event can occur to the number of possible outcomes

outcomes
the possible results of a probability event

P(event) means the probability of an event occurring.

Example 1

Decide whether spinning an odd number is impossible, unlikely, equally likely, likely, or certain.

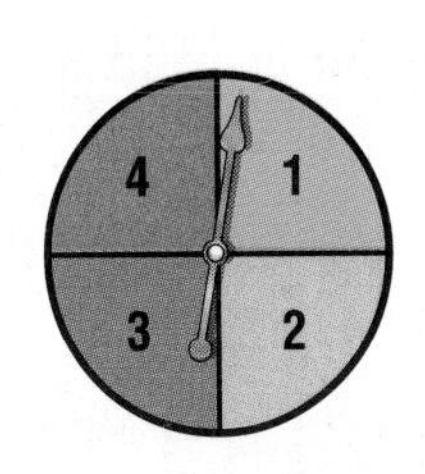

1. On the spinner there are 4 equally likely outcomes.

   1, 2, 3, 4

2. The favorable outcomes are 1 and 3.

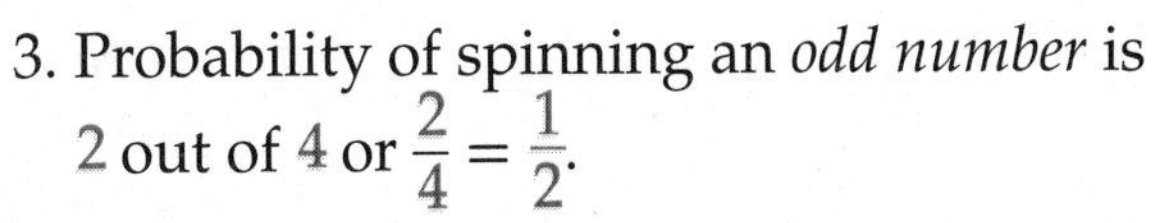

3. Probability of spinning an *odd number* is 2 out of 4 or $\frac{2}{4} = \frac{1}{2}$.

4. The probability of spinning an odd number and an even number is equally likely.

GO ON

YOUR TURN!

Decide whether spinning a 2 or less is impossible, unlikely, equally likely, likely, or certain.

1. On the spinner there are ______ equally likely outcomes.

______, ______, ______, ______, ______, ______

2. The favorable outcomes are ______ and ______.

3. Probability of spinning a *2 or less* is

______ out of ______ or $\frac{\square}{\square} = \frac{\square}{\square}$.

4. The probability of spinning a 2 or less is ______________.

Example 2

In a bag of 3 red, 2 blue, 5 green, and 2 black marbles, find the *P*(blue).

1. The total possible outcomes are the number of marbles.

$3 + 2 + 5 + 2 = 12$

2. The favorable outcomes are blue marbles. The number of blue marbles is 2.

3. Write a fraction with the number of favorable outcomes over the number of possible outcomes. Simplify.

$P(\text{blue}) = \frac{2}{12} = \frac{1}{6}$

YOUR TURN!

In a box of 4 yellow, 5 red, 3 purple, and 3 pink markers, find *P*(purple).

1. The total possible outcomes are the

______________.

___ + ___ + ___ + ___ = ___

2. The favorable outcomes are

______________.

The number of

______________ is ___.

3. Write a fraction with the number of favorable outcomes over the number of possible outcomes. Simplify.

$P(\text{purple}) = \frac{\square}{\square} = \frac{\square}{\square}$

Guided Practice

Decide whether each event is impossible, unlikely, equally likely, likely, or certain when rolling a number cube.

1. roll an even number

   There are ____ equally likely outcomes.

   There are ____ even numbers.

   So, there are ____ favorable outcomes.

   Probability of rolling an even number is

   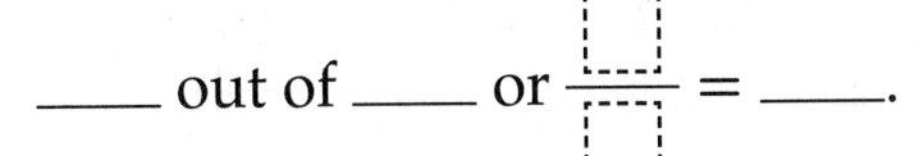

   ____ out of ____ or $\frac{\square}{\square}$ = ____.

   The probability of rolling an even number and an odd number is

   ________________.

2. roll a 4

   There is ________ 4.

   So, there is ____ favorable outcome.

   Probability of rolling a number 4 is ____ out of ____ or $\frac{\square}{\square}$.

   The probability of rolling a number 4 is ________________.

3. roll a number greater than 2

   There are ____ numbers greater than 2. So, there are ____

   favorable outcomes.

   Probability of rolling a number greater than 2 is ____ out of

   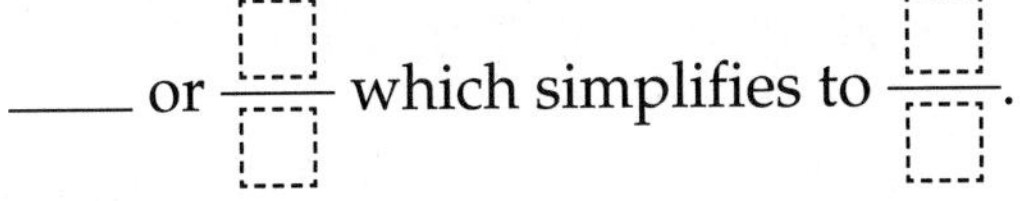

   ____ or $\frac{\square}{\square}$ which simplifies to $\frac{\square}{\square}$.

   The probability of rolling a number greater than 2 is

   ________________.

Step by Step Practice

4 In a bag of 3 red, 4 green, 2 white, and 3 yellow game pieces, find the $P(\text{green})$.

Step 1 The total possible outcomes are the number of game pieces.

_______ + _______ + _______ + _______ = _______

Step 2 The favorable outcomes are the number of _______ game pieces. There are _______ green game pieces.

Step 3 Write a fraction with the number of favorable outcomes over the number of possible outcomes. Simplify.

$P(\text{green}) = \frac{\square}{\square} =$ _______

Find the probability of each event using a calendar of the month of June. Write the probability as a fraction in simplest form.

5 $P(\text{an even numbered date}) = \frac{\square}{\square} =$ _______

6 $P(7 \text{ or } 10) = \frac{\square}{\square} =$ _______

Step by Step Problem-Solving Practice

Solve.

7 **PEP RALLY** There are 13 boys and 10 girls in a class. Each student's name is placed in a hat and one name is randomly drawn to represent the class at the next pep rally. What is the probability that the name drawn will be a girl?

Check off each step.

_____ **Understand: I circled key words.**

_____ **Plan: To solve the problem, I will** _______________.

_____ **Solve: The answer is** _______________.

_____ **Check: I checked my answer by** _______________.

Skills, Concepts, and Problem Solving

Use the spinner at the right to determine if the given event is impossible, unlikely, equally likely, likely, or certain.

8 *P*(odd numbers) ______________

9 *P*(5) ______________

10 *P*(prime numbers) ______________

11 *P*(0) ______________

ALPHABET The letters of the alphabet are written on separate pieces of paper and placed into a bag. Find the probability of each event occurring if a piece of paper is randomly drawn from the bag of the entire alphabet. Write the probability as a fraction in simplest form.

12 *P*(H) = ________

13 *P*(vowel) = ________

14 *P*(M or N) = ________

15 *P*(consonant) = ________

EDUCATION The table shows Hillcrest School District's high schools and their ninth grade teachers. Assume that students are randomly assigned to both schools and teachers. Write the probability in simplest form.

School	First Grade Teachers
Anderson High School	Akers, Frank, Simms
Hudson High School	Latscha, Zimmer
Ludlow High School	Francisco, Melton, Yu
Parsons High School	Kennedy, Peters, Sweeney, Tillman
Silverback High School	Baker, Uhlhorn
Volton High School	Perry

16 Scott is entering ninth grade in the Hillcrest School District. What is the probability he will have a teacher from Silverback High School?

____

GO ON

17 Jasmine is attending ninth grade at Parsons High School in the fall. What is the probability she will have Mr. Kennedy as her teacher?

____

18 Felipe has moved into the Hillcrest School District for his ninth grade year. What is the probability he will attend Parsons or Volton?

____

Vocabulary Check **Write the vocabulary word that completes each sentence.**

19 The ratio of the number of ways an event can occur to the number of possible outcomes is __________.

20 A(n) __________ is the possible result of a probability event.

21 A type of outcome is a(n) __________.

22 **Reflect** Describe an event's likelihood of occurring compared to percentages.

impossible, unlikely, equally likely, likely, or certain

____

____

____

STOP

Progress Check 3 (Lessons 6-5, 6-6, and 6-7)

Use the Fundamental Counting Principle to find the number of possible outcomes.

1 Choose a vowel from the word *magnet* and then choose a consonant from the word *refrigerator*.

____

2 Choose between the fruits banana, apple, strawberry, and orange and then choose between the vegetables carrot, tomato, and cucumber.

____

DRAWING The students' names are Kurt, Lara, Tonya, Jamal, Jacob, Diana, Abey, Charlotte, Tessa, and Hugo. Find the probability of each event occurring if a piece of paper is randomly drawn from the bag. Write the probability as a fraction in simplest form.

3 P(Hugo) $\frac{\square}{\square}$

4 P(girl) $\frac{\square}{\square}$

5 P(boy) $\frac{\square}{\square}$

6 P(begins with T) $\frac{\square}{\square}$

Use the following to solve Exercises 7–9.

7 How many sundae combination choices are available?

____

Sundae Choices		
Ice Cream Flavors	Syrup	Toppings
Vanilla	Strawberry	Sprinkles
Chocolate	Butterscotch	Nuts
Strawberry	Hot fudge	Cherry
	Marshmallow	

8 Suppose the ice cream shop runs out of sprinkles. How many choices are now available?

____

9 If another ice cream flavor is added to the menu, how many choices would then be available?

____

Chapter 6

Chapter 6 Test

The data in the line graph shows the average price of a dozen eggs in January 2003 to 2009. Use this graph to answer Questions 1 and 2.

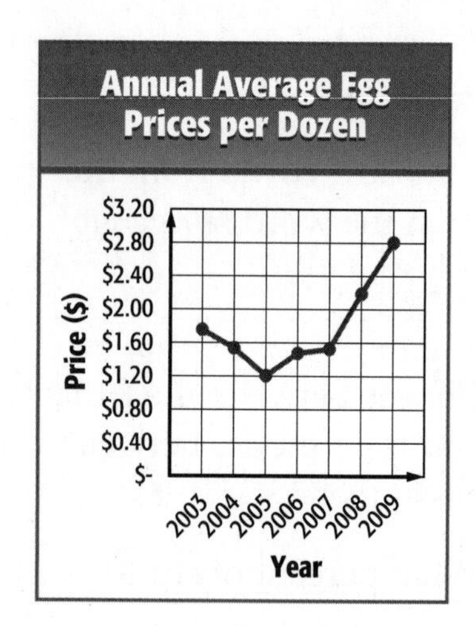

1 Between which years was the *greatest* decrease?

____

2 What was the price of a dozen eggs in 2007?

____

The histogram shows the number of books in 20 library branches. Use this histogram to answer Questions 3 and 4.

Books at Local Libraries

	5–9	10–14	15–19	20–24	25–29	30+
Frequency	2	5	8	3	1	1

Number of Books (thousands)

3 How many libaries have 20,000 or more books?

____

4 In which interval(s) do the least number of libraries fall?

____

The data in the table shows the sports played by a sixth grade class. Create a bar graph to display the data. Use the graph to answer Questions 5–7.

Sport	Number of Students
softball	13
basketball	24
swimming	15
soccer	27
gymnastics	9
football	11

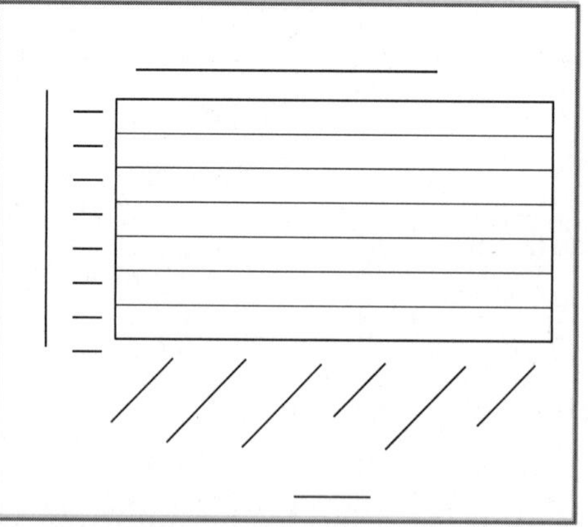

5 What is the difference in the number of students who play the most popular and the least popular sport? ______________________

6 About how many students play football? ______________________

7 How many students were surveyed? ______________________

Use the data in the table to answer Questions 8–12.

Favorite Music	
Popular	60
Hip-Hop	45
Rap	30
Alternative	15

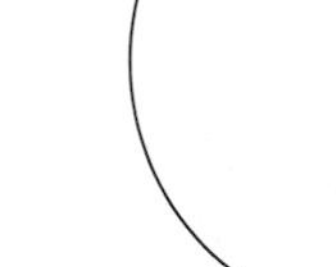

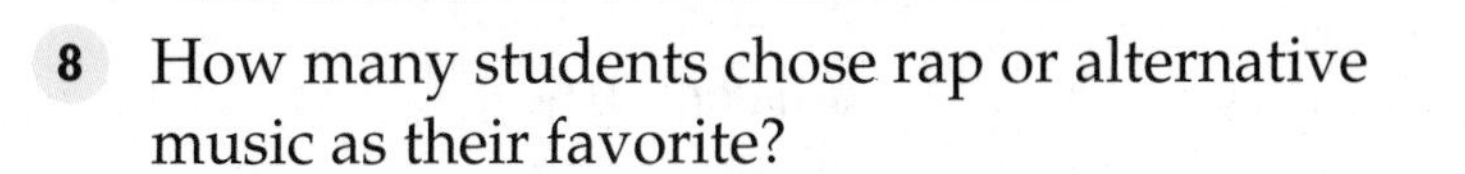

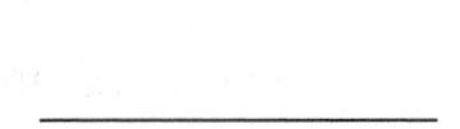

8 How many students chose rap or alternative music as their favorite? ____________

9 What percentage of students like popular music? ____________

10 What percent of students like hip-hop? ____________

11 Create a circle graph using the data in the above table.

Find the mean, median, and mode for each data set. Round to the nearest tenth if needed.

Data Set	Mean	Median	Mode
41, 52, 47, 45, 51, 48	**12** ________	**13** ________	**14** ________
9, 6, 7, 6, 8, 1, 3	**15** ________	**16** ________	**17** ________

A bag of marbles contains 11 green, 8 purple, 4 yellow, 6 gray, and 3 black marbles. Find the probability of each event occurring if a marble is randomly drawn from the bag. Write the probability as a fraction in simplest form.

18 $P(\text{yellow}) = \frac{\square}{\square}$

19 $P(\text{not green}) = \frac{\square}{\square}$

20 $P(\text{purple or black}) = \frac{\square}{\square}$

21 $P(\text{yellow or gray}) = \frac{\square}{\square}$

Correct the mistake.

22 From the choice in the table, Antonio said that he could make 24 different outfits. Is he correct? If not, explain his mistake and give the number of possible outfits.

____

____

____

Outfits		
Pants	Shirt	Shoes
brown	pink	brown
blue	white	black
white	yellow	
black	lavender	

Index

G

S

T

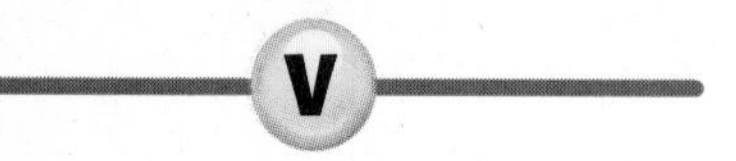

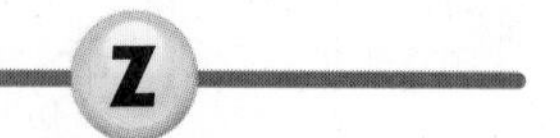